# Brainstorming for a Beautiful Planet

***2nd edition***

*By*

Nicholas Varias

First published on Amazon (Kindle Edition)
August 17, 2022

# Table Of Contents

# About the Author

Nicholas Varias is an award-winning Canadian architect, urbanist, inventor, building code official, and artist. A passionate nature lover, he believes in a radical change of architecture and urban planning and in outer space expansion to preserve the Earth.

Nicholas was born in Arad, Transylvania, Romania, in 1948. He graduated from the "Ion Mincu" University of Architecture and Urban Planning, Bucharest, Romania, in 1972, at the top of his class. After two years of working in Bucharest and two years in Israel and Greece, Nicholas came to Canada in 1976. From 1976 to 1993, he was employed by several architectural firms in Montreal, Toronto, and London, Ontario.

Between 1993 and 2010, Nicholas's architectural practice has successfully completed projects for Canada Mortgage and Housing Corporation, Public Works & Government Services Canada, Department of National Defence, ProFac Facilities Management, and private individuals and corporations.

"Cita Vitae" in Cairo has been an exquisite multi-purpose project for which Nicholas visited Egypt in 1998. In 2002 & 2003, in Barrie, Ontario, Nicholas's firm was involved in several downtown revitalization projects, including the emergency shelter for the homeless, and designed the "Am Shalom" Synagogue.

In 1996, Nicholas won the Canada Mortgage and Housing Corporation's FlexHousing Design Competition; a model unit of his design was built at the Canadian Centre for Housing Technology in Ottawa, for which he had been the design architect. A "Story of Love and Success," created by Nicholas, and published by CMHC, gives an insight into Nicholas's vision for his Flexhouse™.

Over ten years of research and design efforts have resulted in Flexsola™, which is the environmental upgrade of Nicholas' Flexhouse™ design. Flexsola™ is based on Nicholas's patent CA 2843797 for a modular building system. It has received the 2017 Toronto Construction Association's s Innovative Product Award.

During forty years of working as an architect, Nicholas has gained extensive knowledge and experience with the Ontario Building Code, and since 2002, he has been working for several municipalities in Ontario.

Nicholas enjoys sketching and painting, he is a passionate chess player and nature lover and likes the arts, jazz, and classical music. In addition to English, Nicholas also speaks French, Romanian, and Hungarian.

# Synopsis

### Foreword

This book presents ideas for a major reset in architecture and urbanism, which could ensure Humankind's long-term sustainability on Earth in harmony with Nature and adapt to climate change. These ideas aim to elevate architecture and urbanism to a higher level by creating a new paradigm for a future built environment. The book also introduces innovative concepts for extraterrestrial habitation, which would help the Earth recover, and would ensure the survival of humanity in outer space.

### Chapter 1: Flexsola™

This chapter describes in detail the Flexsola™ modular building system, which is based on the author's patent CA2843797. At its core is the unique saddle connection, which permits the fastening of structural columns and beams with bolts only, providing unique flexibility. The fabrication and testing of a prototype module are also illustrated.

### Chapter 2: The Flexsola House

This chapter introduces the design of a model Flexsola House and provides the background and detailed information. The Flexsola House uses the Flexsola™ modular building system described in Chapter 1.

### Chapter 3: Flexsola City

This chapter describes the author's criteria for designing sustainable developments, which include equitable land ownership and rights to use, self-sufficiency, local waste treatment, local food production, sun and light exposure, superior economics, and long-term maintenance. This chapter also introduces the author's concept of Elevated Urbanism to end the anarchy of urban concentration, which has created the multitude of current societal problems we face today. The author also provides the design for an elevated urban community using the Flexsola™ modular system.

### Chapter 4: The Honeycomb

This chapter provides the design background and criteria for the honeycomb modular system and describes the design of an elevated honeycomb house. Details of the author's patent application for a hexagonal modular system are also provided.

### Chapter 5: Honeycomb City

This chapter continues the concept of Elevated Urbanism from Chapter 3, and provides two examples of elevated urban community concepts that use honeycomb modules described in Chapter 4.

### Chapter 6: Honeycomb in Space

This chapter describes the author's architectural designs for creating space habitats with artificial gravity, constructed with honeycomb modules.

### Epilogue

The image of a flower breaking through concrete symbolizes the triumph of Nature.

# Foreword

My first argument in this book is the urgent need to reset architectural and urban planning practices, which are essential for the preservation of our planet and to ensure our long-term survival. I believe that this reset could be accomplished by elevating architecture and urban planning both physically and spiritually. The aim is to create environmentally friendly and self-sufficient communities constructed with innovative modular systems and linked by elevated fast trains. Transportation, high-density human habitation, and activities shall take place above the ground with minimal interference with the natural environment.

My second argument is that the existential conflict between human society and the natural world could be resolved only if people could live and work in the immediate vicinity of the earth. This will permit the Earth to recover and be preserved as a museum for future generations to visit from outer space. The effort to promote extraterrestrial human habitat should be a necessary complement and not a substitute for supporting the efforts to curb climate change, which is part of the first argument of this book. The extra-terrestrial expansion would also ensure the survival of humanity when our planet ceases to exist.

The clock is ticking, and there will be no second chances after the Earth's climate reaches the point of no return. Regardless of if one believes in climate change or not, the risk of not doing what the scientists are asking is too great for the entire world! Furthermore, the sun could explode at any time, destroying the entire solar system. Therefore, preparing for a space migration is the only way to save the human species.

Since 2013, when I published my first book "Brainstorming for a Beautiful Planet" [https://www.amazon.ca/Brainstorming-Beautiful-Planet-Nicholas-Varias-ebook/dp/B00CABM6A4], a lot of events took place, which made me think that a sequel would be a good idea.

The recent Covid 19 pandemic has precipitated my decision to write this sequel, and in a rather peculiar way, it has facilitated its inception. Over two years of frustrations generated by the restrictions to my freedom of movement produced by this unique period in modern history have found a great cure. I plunged into the sea of my creativity, forgetting about my frustrations. This is how many of the architectural and planning ideas in this book were born.

The last ten years have brought unprecedented developments in space exploration. They have encouraged me to revisit my ideas for extra-terrestrial habitats from my first book. I have associated these ideas with my work on modular building, which has produced the Flexsola™ and the Honeycomb patents.

The 3STAR concept of 2013 has evolved into Honeycomb in Space. I have also applied these modular systems to the disruptive concept of elevated architecture and urbanism.

I call it disruptive because it requires a fundamental rethinking of how land is used to eliminate the misuse of the past. I also expect strong opposition from many architects, urbanists, real estate developers, and politicians.

Part One of my first book was devoted to brainstorming on socio-economic and political factors which affect our planet, followed by Part Two, which provided architectural and urbanistic concepts.

In this sequel, I decided to blend the socio-economic and political considerations in each chapter. I think it gives a better understanding of how my thoughts and feelings have determined my designs.

I would like to think that my experience as a practicing architect and as a building code official, associated with my love for nature, has given me a balanced perspective on the needs of the public, as opposed to the interests of private individuals and corporations. I strongly believe that these needs can no longer be met at the expense of the natural world.

Having for long struggled with a creative ego common to most architects; I am now using it to address climate change rather than create unique and spectacular buildings. This professional change of heart became my guiding lighthouse for the rest of my career.

By exposing my design ideas, I am aiming to elevate architecture and urbanism to a higher level and to help create a new paradigm for the future built environment. This paradigm combines the use of advanced modular building systems with raising streets and buildings above the ground to reduce the burden on the natural environment.

My ideas were inspired by Nature! She creates life modules, which we call cells. By using modular design, architects are therefore following in Nature's footsteps. We have been doing it for thousands of years with bricks and stones. Furthermore, if one observes how natural ecosystems, such as forests, are organized, a vertical layering of various species of plants and animals becomes apparent. Some trees grow up higher than others, raising their crowns above the ground, which is left for the "little ones" (shrubs, grasses, mushrooms, etc.) to occupy.

What a wonderful natural world we live in, and we have the privilege of sharing it with other life forms! In my view, it is inevitable that the two concepts, modular building, and elevated architecture, must work together.

Ecological modularity, which could achieve building flexibility in space and time, is essential for sustainable development. It is the subject of Chapters 1, 2 & 4. I became interested in modular buildings in 2015 when I came up with the idea of a bolted connection system that could provide the flexibility missing in the current systems.

In 2017, I registered the Canadian patent for Flexsola™, which is a unique modular building system. The designs in Chapters 1, 2 & 3 are based on this patent.

More recently, I have applied for a patent for hexagonal modules, which I used in Chapters 4, 5 & 6.

During my early years as a practicing architect, I created several concepts of buildings raised above the ground. These can be considered precursory to my current ideas for elevated architecture and urbanism, which are presented in Chapters 3 & 5.

A first example is my proposal for a **Tall Ship Dinner Theatre** in Port Stanley, Ontario (F-01). The building would be placed close to the beach, with the bow facing the lake and the ground floor reduced to a minimum required by a glazed lobby, receiving & service rooms, and vertical circulation.

F-01

A more recent concept for an **Art and Science Centre** (F-02) is an attempt to reconcile Art, Architecture, and the Environment, by raising the building on piers.

By suggesting two hands holding a globe of light, the concept symbolizes human values and achievements, such as technological progress, caring for the environment, community spirit, and partnership.

Functionally, the "palms" and "fingers" accommodate a combination of public and private uses such as art galleries, performing arts, science centres, mini hôtel, boutiques, and restaurants. The “wrists” are reduced to the minimum required by structural piers, stairs, and elevators to minimize the obstruction of the views to and from the water, signaling my desire to elevate architecture.

The glazed geo-sphere would shelter a mini ecosystem for advanced scientific research and experimentation, including solar energy. It would also provide a year-round flower and butterfly garden.

Added as a science fiction element, the "flying saucer" with vertical uplift would reinforce the high-tech appearance, and it could also be developed as direct access into the geo-sphere. On the other hand, the organic shapes are intended to achieve harmony with nature and the surrounding landscape.

Many years ago, I expressed my feelings about the world in a series of three symbolic paintings, which I called **Triptico**. They are accompanied by my verses, which facilitate the understanding of the images. Triptico was first published in the 1$^{st}$ edition of my book in 2013. Since then, my feelings and thoughts have not changed; therefore, I decided to include them in this book.

## *Nature Crushed*

The merciless tank of destruction,

Crushing all nature's creations

Day and night, a chain of divers,

Carry petrol into its bowels.

A swastika is running the engine,

And is so big and strong and mean,

Like anything that anyone has ever seen.

Teutonic bull horns are stabbing the sky,

The demonic brain is going high,

The decayed tooth is launching a rocket.

A golden sickle with the almighty handle,

It will soon finish the bloody harvest.

All flora and fauna are fleeing,

With no chance of escaping,

So what?

Politicians still argue and argue.

***The Little Boy and His Teddy Bear***

Suspended in the universe,
By an everlasting curse,
The little boy watches the world,
Feeling abandoned, feeling so cold,
Things appear upside down,
As if played with, by a clown.
Up there flies a deadly bird,
Is it a bomber, is it the third?
After the first two hit Japan,
This may be destined for Iran.
Another ice age may begin,
For cleansing mankind of all sin.
The little boy sucks on the nipple
Of Mother Earth, the poor cripple.
A flying saucer is a tear of sorrow.
He likely sheds for lost tomorrow.
With curling hair as question marks
His look's becoming angry sparks.
How could I rescue this world?
Those humans cannot behold!
Is this the ending of the world?

**The Web of History**

The heavy canon of War,
Is climbing the bloody
Web of History,
All in the name of
Religion and Ideology?
Star of David,
Cross and Crescent,
Red and Blue Stars,
Are all firmly implanted in its
path.
A Butterfly of innocent nature,
And a Dove of Peace,
Are silent witnesses.
But what is there hiding,
In the eye of the web?
The claws of the Spider,
Waiting for the end of history.

The existential conflict between human society and the natural world could be resolved only by immediately planning for people to live and work in the immediate vicinity of the earth. This will permit the Earth to recover and be preserved as a museum. The efforts to promote extra-terrestrial human habitat are a necessary complement and not a replacement for the efforts to curb climate change.

A raised awareness of the essential purpose of space exploration for the preservation of our planet shall bring hope to the millions of our young who currently live without it and consequently have found refuge in cheap momentary gratification. Hope will bring interest and participation, and the contribution of millions will make human expansion in space possible to save humankind and the Earth which sustains us.

In Chapter 6 – Honeycomb in Space, I show how the Honeycomb modular system could be used to develop viable and flexible space habitats.

**Proposed Audience**

- All those who care about the Earth and are interested in innovation.
- Architects, engineers, and those who teach and study these professions. It could inspire them to achieve professional reconciliation with the natural environment and to contribute to effective action for adapting to climate change.
- Progressive real estate developers, who are searching for new opportunities, may find my modular designs a good fit for their innovative developments.
- Space physicists and other experts involved in space exploration may be interested in evaluating my architectural concepts for creating space stations with artificial gravity.

## How to use this book

This book contains proprietary information, including architectural and urban planning concepts, designs, and technical details, which are available for commercial use only with the author's written permission explicitly granted.

# Chapter 1: Flexsola™

Flexsola is the result of several years of research and design in modular building systems which could contribute to a circular economy. It can be used to construct many types of buildings of any size and is particularly advantageous when future expansion, layout flexibility, construction expediency, or demountability are desired. Thanks to these advantages, Flexsola is an appropriate building system for affordable housing for communities in the third world and indigenous peoples, expedient developments in harsh climates, military facilities that can be easily relocated, and post-disaster construction.

The word Flexsola is the acronym for Flexible + Solar and represents the idea of creating a modular building system, which provides functional flexibility and incorporates solar energy in the design of buildings. I believe that Flexsola is unique among existing modular building systems.

The architectural design options, including layout and facade, are practically limitless, which is an advantage for accommodating municipal planning requirements and for integrating additions to existing building styles and urban fabric.

The unique saddle connection makes two installation options possible and requires less labour than conventional systems due to its *guiding pin* component. One option is to install the prefabricated components on-site, which may be the best solution to suit local conditions. A second option is to pre-assemble the prefabricated components into segments of container-size, transport these to the site, and install them with minimal labour required. There are few, if any, other modular building systems that currently offer both these options.

Flexsola creates space modules, which can be added horizontally or vertically, to accommodate future needs without disturbing the existing interior space. Thanks to the bolts-only connections, all components can be easily and quickly assembled or disassembled. Soon, this will be performed with robotic drones, thus further reducing the installation time and labour required on site. This capability reduces costs and facilitates construction in harsh climates. Furthermore, the saddles can be mass-produced with 3D printing, further reducing costs.

With the ability to incorporate solar panels and mini wind turbines without affecting the rooftop, a Flexsola building could produce more energy than it would consume.

The ecological footprint of the building is substantially reduced by a landscaped roof terrace, which returns to nature a large portion of the ground area used. Complemented by an optional greenhouse, this rooftop terrace could be an outdoor oasis for leisure, pets, or growing organic fruits and vegetables in a private and protected environment. It can also function as a collector of rain and snow water.

Current conventional connection methods between the structural steel posts and steel beams are inflexible, do not facilitate prefabrication, and are difficult and expensive to modify when adaptation, expansion, or disassembly is needed. Flexsola solves this problem by introducing a novel connection system that facilitates prefabrication, helps achieve flexible growth, adaptation, or relocation, and can create building modules that function ecologically as "living" architectural organisms. These modules, like living cells in the natural world, can create building blocks that can grow, shrink, or be transplanted. Furthermore, Flexsola facilitates the incorporation of solar panels and mini wind turbines to achieve energy self-sufficiency for buildings.

Currently, there are other modular building systems in use. However, Flexsola is uniquely novel by introducing a specially configured saddle to connect or disconnect posts and beams. Thus, independent, stand-alone modules can be created, which can be added, removed, or relocated within a three-dimensional grid with minimum manpower and equipment and without disturbing the structure of adjacent modules (C1-01).

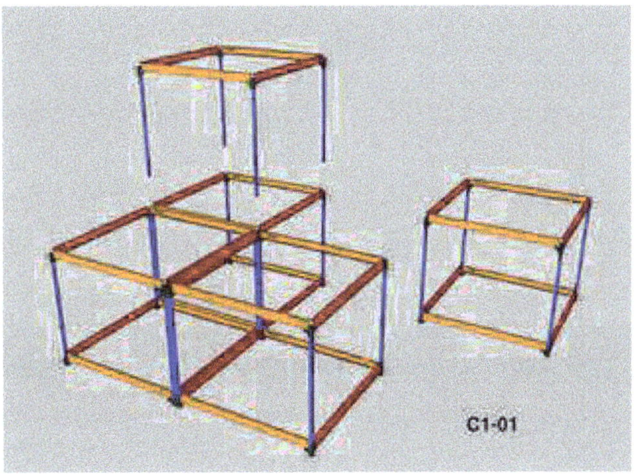

**My case against using wood as a structural material.**

A few years ago, the Ministry of Municipal Affairs and Housing amended the Ontario Building Code to allow the use of wood construction in apartment buildings of up to six storeys [https://www.buildingcode.online/170.html], despite strong objections from numerous municipal fire and building officials, concerned about the safety implications.

I am convinced this change was a result of pressure from the lumber industry and of misguided government policy of facilitating northern development based on the exploitation of natural resources, especially timber. From an environmental point of view, this policy is not different from the federal government's support for the petroleum and lumber industries. This policy continues the two-hundred-year-old colonial practice of exporting natural resources, which has prevented Canada from becoming a major industrial power.

The lumber industry has been promoting wood in construction by claiming its environmental advantage over concrete and steel. In my view, this belief is wrong, as it is based on distorted and incomplete facts. Just by looking at the recent upsurge of clear-cutting in some countries, such as Brazil or Romania, triggered by an increase in worldwide demand for lumber, one can realize the catastrophic effects caused to the global environment.

The lumber industry and its supporters would like us to believe that harvesting the forest is better for the environment than leaving it alone. Furthermore, they claim that, as the trees grow older, they start to decay and release the stored carbon. It is a misleading way of trying to convince us that because it is a recyclable natural resource, wood is the most environmentally friendly building material.

In fact, as recent studies show, instead of slowing down as the centuries add up, old trees speed up their growth.[ https://www.livescience.com/42578-oldest-trees-grow-fastest.html ]

Old forests fix less carbon than new forests, but not at the individual tree level. In fact, larger and older trees accumulate carbon more rapidly than younger ones. The fast growth of older trees means that they have an important role in the global climate system.

The lumber industry, and the government that supports it, are silent about the disruptive effect tree harvesting has on forests and wildlife.

In the following Table, I make a comparison between Concrete & Steel vs. Wood.

| CHARACTERISTICS | CONCRETE & | WOOD |
|---|---|---|
| Non-combustible | YES | NO |
| Require less or no fire protection | YES | NO |
| Do not require water supply for sprinklers | YES | NO |
| Safer during storms or earthquakes | YES | NO |
| Have thermal mass for superior energy performance | YES | NO |
| Require less mechanical equipment for A/C | YES | NO |
| Recyclable | YES | NO |
| Better long-term structural strength | YES | NO |
| Better shape stability | YES | NO |
| Better durability | YES | NO |
| Demountable & reusable | YES | NO |
| Resistance to moisture | YES | NO |
| Resistance to mold | YES | NO |
| Resistance to insects | YES | NO |
| No indoor emissions from adhesives | YES | NO |
| Do not require cutting trees | YES | NO |
| Lower long-term maintenance cost | YES | NO |
| Suitable for modular building systems | YES (1) | NO (1) |
| Suitable for 3D Printing | YES (2) | NO (2) |
| Lower initial cost | NO (3) | YES (3) |
| Less energy to produce | NO (4) | YES (4) |

**Note 1**: For technical reasons, long-lasting modular construction, which is also flexible, is not feasible with construction lumber.

**Note 2**: New 3D Printing technologies are emerging, which are replacing conventional concrete and steel for manufacturing building components without compromising the listed characteristics.

**Note 3**: The lower initial cost of construction lumber is due to being traditionally an almost free resource, easily available to landowners, who could sell it to builders. On the other hand, producing steel or concrete became more expensive, especially after emission controls were introduced. However, the balance has been tilting away from wood in favour of concrete and steel since stricter environmental forestry methods have been introduced, and cleaner concrete and steel technologies are being developed.

**Note 4**: It is debatable which construction material uses less energy, as this depends on several factors, such as location and energy sources.

Furthermore, current costing practices promoted by the lumber industry, which indicate that wood uses less energy, do not account for "hidden" cost factors, such as reforestation and recycling. Not to mention the damage that tree harvesting does to forest ecosystems.

Due to structural performance requirements, the core component of the Flexsola system, the durable connection saddle, could not be mass-produced with wood.

The saddle mock-up (C1-08) was fabricated from steel plates welded together into several components and has been designed for a house. However, similar saddle configurations can be designed for other occupancies with higher loads, as described in Chapter 3.

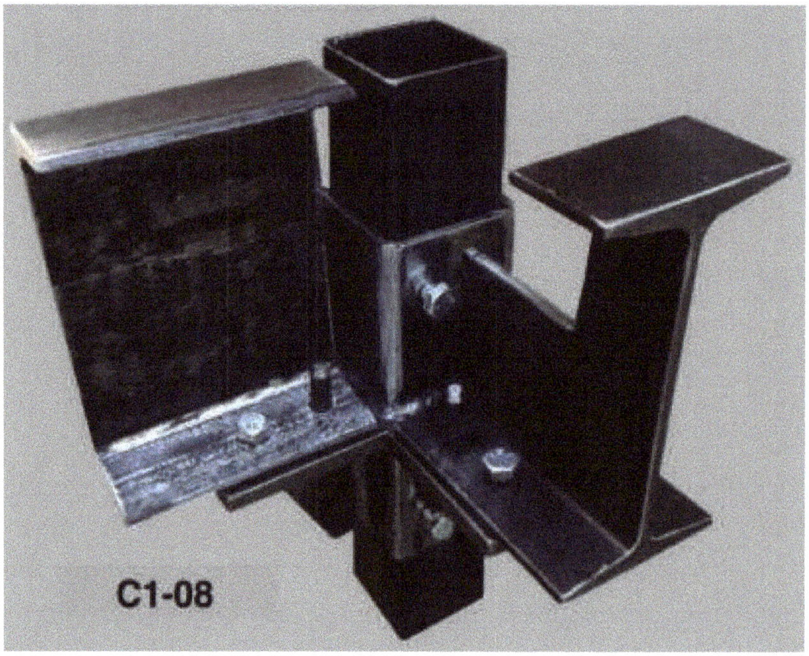

The following drawings and description are taken from my international patent PCT/CA2016051507, which amends my Canadian patent CA2843797 (C1-02, C1-04, C1-05, C1-06 & C1-07).

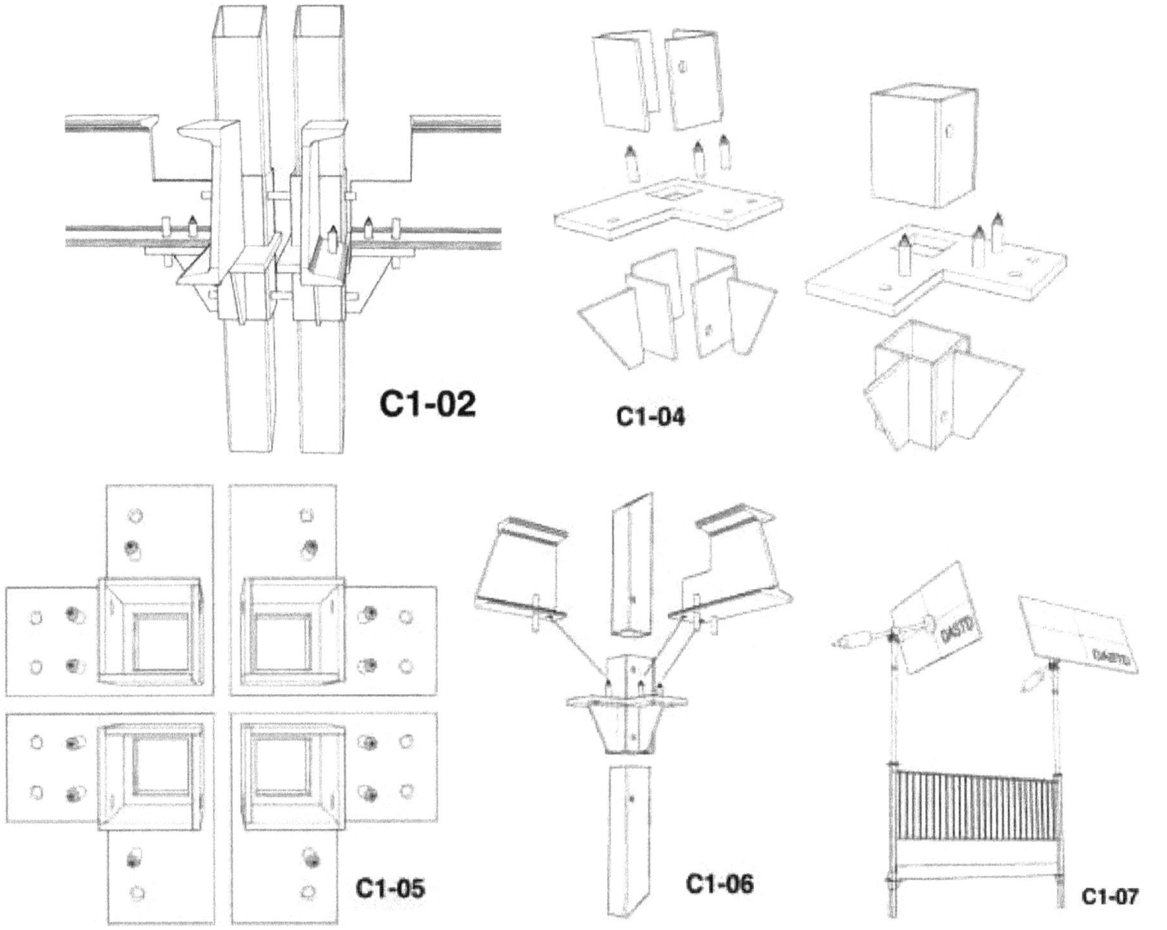


An L-shaped horizontal plate is provided with holes and upward pin/bolts welded to it, which serve to guide the beams into place. After placing the beams, these pins/bolts receive nuts and become fasteners. In the prototype shown here, the plate is 13mm thick and has an overall dimension of 200mm x 200mm, with two equal branches of 100mm x 100mm. The plate has an opening of 64mm x 64mm centered on the axis of the posts to permit the vertical transfer of cables between floors.

The downward and the upward vertical sleeves, 89mm x 89mm x 6mm thick, 114mm high, telescope into the lower and upper posts. Each sleeve is fabricated from four plates welded together to form a tube, which can fit into the post, and has holes for bolting.

Using HSS instead of welded plates is a possibility depending on the availability of adequately sized HSS posts, which could provide a clear interior space to fit the sleeve with little shimming or grinding needed.

Two triangular-shaped downward vertical stiffeners, 114mm x 20mm x 6mm thick, are welded to the horizontal plate in the middle of the plate branches and sleeves.

Following a lengthy design process, I had the saddle prototype and the module assembly (C1-09, C1-10 & C1-11) made in a steel fabricator's shop in Oakville, Ontario. The fabrication took place during the Covid19 pandemic when there was a shortage of steel beams. Therefore, I used rough-sawn solid timber instead.

Two technical challenges arose and have been overcome during testing. Firstly, the selection of the material for fabricating the lower and upper sleeves of the saddle to achieve the telescopic slide of the posts into the sleeves.

Because of tolerances, the next sizes of standard HSS posts left either too much gap requiring shimming or not enough room for grinding. Secondly, proper welding requires a minimum of 6mm thick HSS walls, which results in thicker HSS than required by structural design for vertical loads.

The saddle incorporates a pin, which serves as a guide and keeps the beams in place before the bolts are installed. It facilitates the installation by reducing the labour and equipment required.

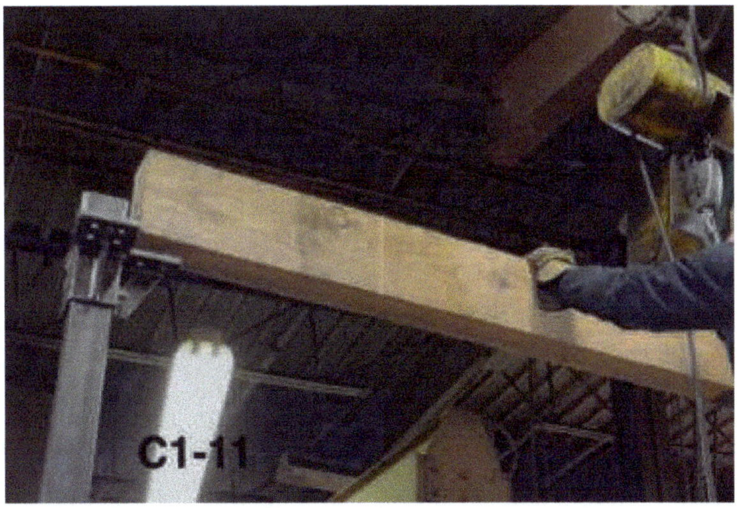


See videos: https://vimeo.com/545167281 ; https://vimeo.com/530990855

To reach the commercial launch, Flexsola requires further testing of the saddle in a real-life setting of multiple structural modules. Furthermore, to achieve its full advantage, Flexsola needs research and development of robotic drones for installing or demounting the saddle connections at higher levels without having to use scaffolding. In addition, the manufacturing and mass production of saddles with 3D printing with metallic powder will greatly reduce manufacturing costs.

**Parapet post extension**

The saddle connectors used at the roof parapet can accommodate parapet post extensions (C1-07), which can be incorporated during construction or after completion. In addition to supporting the guardrails around the rooftop terrace, these parapet posts can also support dual-axis solar tracking devices (DASTD) with PV solar panels, mini wind turbines, or communication devices. All these are placed overhead and do not interfere with the use of the rooftop terrace.

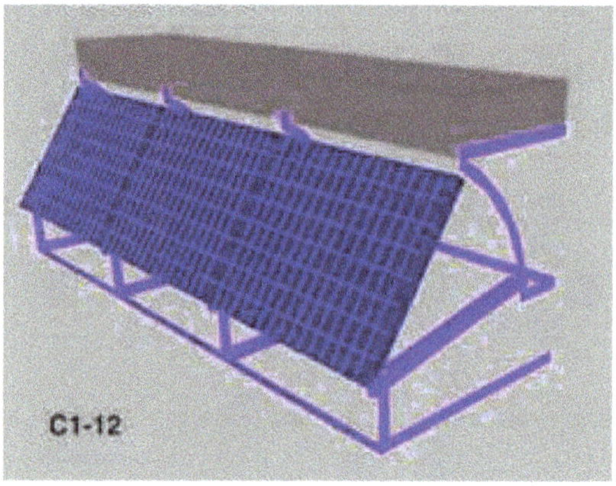


The parapet posts can also support adjustable brackets for solar panels (C1-12), which could also serve as awnings to protect the windows. Furthermore, planters can be attached to the parapet posts. All these features are presented in Chapter 2: The Flexsola House.

**Dual Axis Solar Tracking Device (DAST)**

A Dual Axis Solar Tracking Device (DAST) (C1-13) has four photovoltaic panels mounted on a metal frame, supported by the roof parapet, and elevated above the rooftop terrace. It can produce approx. 1Kw of electricity. The number of units required depends on lifestyle and habits, and additional units may be easily added in the future.

It can be mounted on the parapet at locations that would not cause shading of the rooftop garden. The steel pivoting support with counterweight is designed for easy adjustment to solar altitude and azimuth. The number of DASTDs is subject to the owner's budget and the electricity to be generated for personal consumption or to be sold to the grid. The system has the great advantage of flexibility of location and production efficiency over conventional solar panels mounted on pitched roofs.

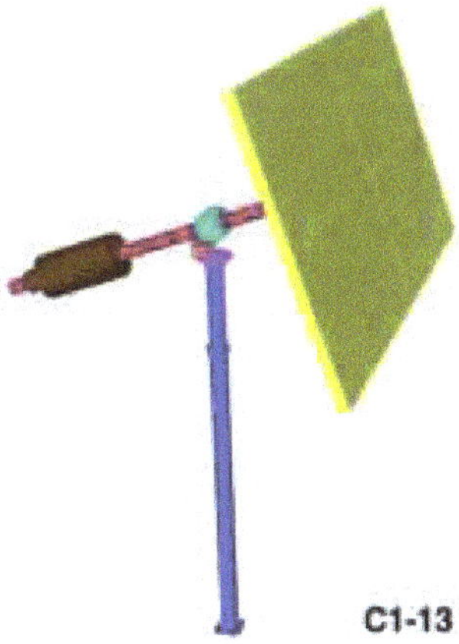

C1-13

**Benefits**

Flexsola has many advantages over existing modular building systems, but the most important is building faster, safer, and more economically, especially in harsh climates and post-disaster or war-torn areas. Furthermore, Flexsola provides greater building flexibility in space and time.

**Recognition**

This photo (C1-14) shows me receiving the 2017 Toronto Construction Association's Innovative Product Award.

# Chapter 2: The Flexsola House

The Flexsola House is a low-density residential application of the Flexsola™ modular building system described in Chapter 1. If mass-produced, it can be cost-effective for providing high quality self-sufficient, and affordable housing anywhere in the world. The benefits of the Flexsola system make these houses attractive for developing countries, indigenous peoples, housing in harsh climates, post-disaster reconstruction, or military facilities that can be dismantled and relocated.

Due to its functional flexibility, the Flexsola Houses could appeal to a wide range of end users, representing a growing percentage of the population: the seniors (i.e., snowbirds), people with special needs (i.e., home care), first-home buyers, or empty nesters, the self-employed and small businesses. The modular concept allows a variety of ownership and/ or tenancy arrangements that can accommodate customer needs and local planning and zoning by-laws.

Arriving at the Flexsola House concept took me over thirty years of innovative thinking. My interest in flexible housing began in 1989 with the design and construction of two houses at 46 & 48 Cliftonvale Ave in London, Ontario (C2-01). The walkout basement had a separate side entrance and could function as an independent one-bedroom apartment.

C2-01

My family and I lived at 46 Cliftonvale Ave (C2-01) until 2000, and my wife used the walkout basement for a home day care to supplement our income and provide valuable social interaction for our children.

A new architectural concept and model (C2-02) followed in 1993-94 when I started my architectural practice in London, Ontario.

This experience has built the conceptual foundation for the FlexHouse™ design, which won the Canada Mortgage and Housing Corporation's FlexHousing Design Competition of 1995-96.

My submission included two model units, "A" & "B", with an identical ground floor (C2-03, C2-04, C2-05, C2-06, C2-7 & C2-08).

C2-03
A-1

C2-04
AA-2

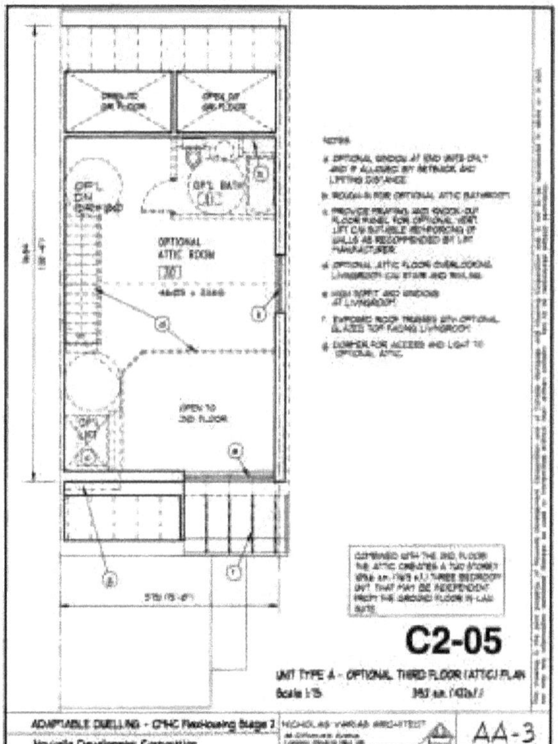
C2-05
AA-3

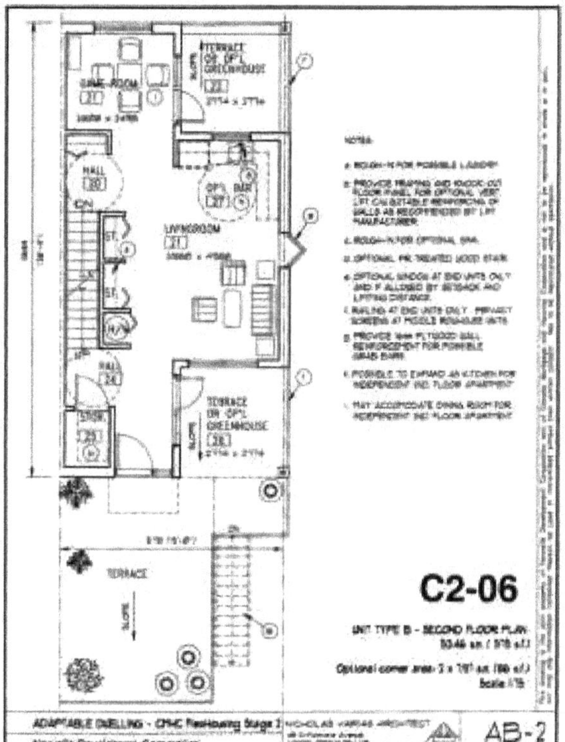
C2-06
AB-2

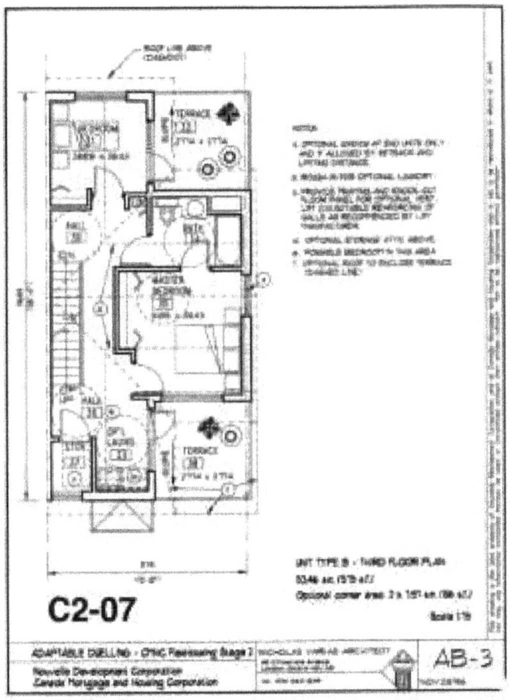

## FlexHouse™ specifications

The floor plans show the unique features of the design. Storage rooms vertically lined up on all levels, have knockout floor panels which permit easy installation of wheelchair lifts if desired.

All interior partitions are non-load bearing and may be relocated. All floors can be made accessible to wheelchair users, and they could be connected by stair lifts or a vertical lift. The owner can use the ground floor as an independent one-bedroom in-law suite and/or home office.

In locations that allow legal duplexes, construction may be easily modified to achieve full separation of the ground floor from the top floors and meet the safety requirements of building and fire codes.

Potential future expansion is achieved at unit type "A" by adding a 3rd-floor attic space within the same building envelope and at unit type "B" by enclosing any of four corner terraces.

Economic passive solar heating may be achieved by adding greenhouse(s), distribution ducts through open floor space, connected to H.V.A.C. ducts, and fans. The use of prefabricated space joists and T-bar ceiling allows for easy installation and future modification of mechanical and electrical systems.

Solar heat is to be stored in a crushed stone bed under a slab on grade and optional concrete floor topping; alternatively, non-combustible concrete floors and walls would provide thermal mass.

A model of my winning design was built in 2006 at the Canadian Centre for Housing Technology in Ottawa (C2-08a, C2-09), unfortunately with bad modifications to the attic, which I have not agreed with.

C2-08a

Canadian Centre for Housing Technology - News - Issue 2

**Meet FlexHouse architect Nicholas Varias**

Nicholas Varias, an architect from London, Ontario, designed the FlexHouse that won Canada Mortgage and Housing Corporation's 1996 FlexHousing Design Competition.

The challenge was to design a multi-purpose home that, with very little effort, could change according to the family needs. This house is on display at the CCHT's InfoCentre in Ottawa as an example of the best in housing innovation and technology.

Varias has been living since 1990 in a home that he designed and built with FlexHousing features. The lower floor is a self-contained unit, once occupied by his wife's day-care business and now a guest suite for visiting relatives. The unit could be rented out as an apartment.

His design addresses the need for neighborhood spirit and community. "People don't know their neighbours, and the elderly and people with disabilities are sent to live in special places," he says. "In the FlexHouse, everyone has a place in the home, both young and old."

His award-winning design is illustrated in a foldout called Amanda's Story. It shows how a typical home built according to FlexHousing principles can easily be reconfigured to accommodate families in flux. Amanda's Story is available through CMHC's Canadian Housing Information Centre at 700 Montreal Road in Ottawa or by phone at 800-668-2642, by fax 613-748-4069, or e-mail at chic@cmhc-schl.gc.ca

C2-09

The FlexHousing™ concept is made easy to understand by Amanda's story, which I wrote, and which was edited and published by CMHC as "The Circle of Life" in FlexHousing: Homes that Adapt to Life's Changes, in FLEXHOUSING™ HOMES THAT ADAPT TO LIFE'S CHANGES – ISBN 0-660-17477-4. My original story follows.

***A Story of Love and Success***

*In 2015, Amanda Smart was a young teacher, hardworking, intelligent, and open-minded. She purchased a type "A" unit and rented the ground floor to Darren Strong, who was on his way to becoming a lawyer. With the money from Darren's rent and the extra electricity generated by the solar panels, Amanda was able to carry the mortgage and make some additional investments, such as saving for retirement.*

*Amanda and Darren became friends. They shared the rooftop garden and laundry but also discovered they had many common interests, such as enjoying nature, star watching, and karate. In 2016, they married. Darren finished his studies and opened his office downstairs.*

*Many of his clients were physically challenged, and his ground floor office was fully accessible to wheelchairs. Darren sold his car, and because he worked at home, they both shared Amanda's Toyota.*

*A year later, baby Justin came along, and with him, a much-needed new laundry facility on the second floor. Amanda's mother moved in to look after the baby when his parents were working. Justin truly enjoyed spending his first years at home, being raised with loving care. As his parents needed extra room, they added an attic, including a new bedroom overlooking their livingroom. They called it "our little paradise."*

*Six years later, in 2023, grandma became confined to a wheelchair. Rather than assigning her to a nursing home, Amanda and Darren decided she should stay with their small family. A lift was installed to allow grandma free access to all floors.*

*In 2035, grandma passed away, and they all missed her very much. By this time, Darren had moved out from downstairs and was now sharing an office with his partner.*

*Justin, now a teenager, moved downstairs into his own apartment. He enjoyed the freedom of coming home late without disturbing the rest of the family.*

*Within 24 months, he left home for university, and Amanda and Darren found themselves alone. As the house was now too big for them, they decided to rent the ground floor again. They could also use the money to help pay for Justin's tuition fees.*

*Justin brought his bride home in 2043. The newlyweds had no money and needed a place to stay. Like most parents, Amanda and Darren decided to help and let the young couple move into the ground-floor apartment.*

*In 2046, Justin and Alice had a baby. As they needed an extra bedroom, they moved to the second floor. The "little paradise" was large enough for the aging couple. They could also rent the ground floor again. Then, the grandparents moved downstairs, and the growing young family could use the attic as well. "Why move," they reasoned, "when we can live in the same house and be independent at the same time?" And most important, they could all care for each other.*

*And the story of love and success goes on.*

My experience with FlexHousing™ had been a roller coaster. After winning the competition, I had great hopes and expectations for building a model unit in London, Ontario. I have come close to succeeding in 1998. Unfortunately, my agreement with a builder fell through when the developers who owned the lot pulled out of the deal.

After moving to Barrie in 2000, I tried again to develop FlexHousing™ (C2-10), using my own property on Gunn Street, which I had planned to sever in several lots, but I hit the wall of municipal bureaucracy, and my hopes were dashed again.

# Home designer says, "Be flexible"

C2-10

Huronia Business Times - Aug. 2001

Another attempt followed in 2003 in Orillia when a developer contemplated building two semi-detached units. Unfortunately, he, too, got cold feet. Consequently, after moving to Oakville in 2004 to work for the Town of Oakville in the building department, I have decided to concentrate on upgrading my concept and promoting it as Flexsola™.

In retrospective, my lack of success in building a model unit was due to a combination of several reasons. Firstly, I could only blame myself for my lack of business and marketing skills required for such an endeavor. In terms of design, I would like to think that I am a capable and creative architect. Unfortunately, I am not a businessperson too. Furthermore, I am quite outspoken and lack certain social skills needed for marketing. And finally, I made the mistake of not seeking a suitable partner who could compensate for my shortcomings.

Secondly, CMHC had the policy of limiting their support to a publicity campaign, which failed to convince the housing developers and builders to embrace the new idea. After the fifteen years, I had spent trying to promote his FlexHouse™ design. I am now convinced that a little financial support would have made the difference.

It should have been obvious from the start that the financial bottom line is the main motivation for running the housing sector. It would be unrealistic to expect a builder to risk undertaking a new concept, which, although it theoretically makes sense, has not been tested by the market. On the contrary, for many years, homebuilders have been making a large profit with house models repeated to such an extent that one could not distinguish between various Ontario neighbourhoods. Why would they risk this market-proven practice for the sake of helping people move less?! It would be like undercutting their main source of revenue based on the number of homes sold. In addition, real estate brokers would surely agree as well. No wonder I have not met to date any realtor interested in FlexHousing™. Therefore, the third reason is a conflict of interest.

Nevertheless, on the positive side, during these years of trying to build a FlexHousing™ momentum, I met hundreds of individuals of different backgrounds and ages who all loved the concept and indicated they would be interested in owning a unit if it was available.

Yet, there is another important reason that has prevented a breakthrough for the FlexHouse™ concept. Its foundation relies on its flexibility to function as one or as two independent dwellings. The potential of having a secondary unit gives it its strength in competing with conventional housing and is essential for its viability and success. However, this strength proved to be also a weakness, as it conflicted with zoning by-laws that do not allow single-family homes to have a secondary apartment.

There has been a long political battle over the right of homeowners to have a second unit in their homes, even though this could create much-needed affordable housing. Under the growing pressure of the need for affordable housing, many municipalities have recently changed their zoning policies to allow secondary suites, subject to meeting building code requirements.

Following is my article on this subject, published in the "Oakville Beaver" on January 07, 2011:

## *Affordable housing alternative*

*Adequate shelter is one of life's essentials. The continuous rise in house prices due to growing demand and the escalation of costs associated with land, infrastructure, and construction has exacerbated the crisis of affordable housing in Ontario. Not surprisingly, the Ontario government has recently introduced Bill 140, the Strong Communities through the Affordable Housing Act.*

*As part of a long-term affordable housing strategy, this legislation will mandate municipalities to establish policies that authorize the use of a second residential unit in a detached house, semi-detached house, or rowhouse. As indicated by Canada Mortgage and Housing Corporation, secondary units are one of the most cost-effective ways of providing affordable housing.*

*Today, as governments at all levels struggle to reduce their budgetary deficits, it makes sense to rely more on the private sector for creating affordable housing. A house can be designed with the flexibility of creating an independent secondary suite if needed without adding to its cost. This feature is part of the FlexHousing concept, also promoted by CMHC.*

*Second or secondary suites are also called accessory or auxiliary apartments, granny flats, in-law, or guest suites. They can provide suitable accommodation for many individuals, such as students, pensioners, and people with disabilities, which are presently in short supply. By reducing the need for publicly financed affordable housing, they contribute to economic and environmental sustainability, with less use of energy and natural resources.*

*Secondary units facilitate home ownership by providing first-time homebuyers with the option of renting out part of their home. They get flexibility in time and space by using the house as needed without compromising privacy. For example, many young first-time homebuyers do not need or afford more than a small apartment, yet they prefer to live in a larger house.*

*Then, as they establish a family, they can occupy the entire house rather than move. And finally, after their grown-up children leave, the second suite could be rented out again to supplement the pension.*

*By encouraging multi-generational living, second units have positive social ramifications, such as home care for the children and the elderly, preferred in many traditional families. Second, units strengthen a neighbourhood by reducing the need to move due to economic hardship. By renting out the second suite when facing financial difficulties, homeowners can more easily pay the mortgage and keep their homes.*

*For people who are away for longer periods, such as "snowbirds," a second unit can provide security by maintaining an indoor presence and supervision during their absence. At the same time, it helps and provides companionship, especially important for seniors and individuals with disabilities. To some homeowners who need this option, renting out part of their house can be very important, even crucial.*

*Therefore, when considering the pros and cons of allowing them to have a second suite, their property rights shall be given priority.*

*Historically, in many cases, the objections against second suites originated from the NIMBY (not in my backyard) syndrome rather than from an educated analysis. Still, there are legitimate concerns relating to traffic, shortage of parking, potential property neglect, and neighborhood degradation, which would lead to lower real estate values in the area. To address these concerns, a municipality can take various measures to ensure that the functioning and appearance of neighborhoods are not negatively affected.*

*First, convenient public transportation must be provided to reduce the need for parking in compliance with new transportation policies. At the same time, issuing street parking permits could satisfy the short-term needs that may be produced by second suites and would also generate municipal revenue. Secondly, registration and monitoring are essential to ensure that second suites are created and maintained only in owner-occupied homes. Thirdly, a permit application shall be required to review compliance with the building code and municipal standards.*

*The introduction of a regulated mechanism for allowing second suites will substantially reduce the number of illegal accessory apartments, many of them unsafe. Statistically, second suites translate into a small increase in density. Nevertheless, people must accept the fact that population growth is inevitable and so may consider moving to a less densely populated area, but they would have to face other challenges, such as transportation and fewer services.*

*The strengths and security of a neighbourhood rely on the relationship established among its residents. As opposed to gated and low-density communities, which in effect create human ghettoes, in a more populated and mixed community, there seems to be more interaction between neighbours. There is overwhelming evidence, based on the experience of numerous cities, that second units bring much more benefits to urban intensification.*

*Some who prefer to live on a larger property with more privacy and quiet may consider moving to a less densely populated area, but they would have to face other challenges, such as transportation and fewer services. The strengths and security of a neighbourhood rely on the relationship established among its residents. As opposed to gated and low-density communities, which in effect create human ghettoes, in a more populated and mixed community, there seems to be more interaction between neighbours.*

*There is overwhelming evidence, based on the experience of numerous cities, that second units bring much more benefits to a community than problems, problems that can be resolved through wise municipal policies.*

As mentioned in Chapter 1, the word Flexsola™ is the acronym for Flexible Solar and represents the idea of creating a modular building system, which in addition to providing flexibility, also incorporates solar energy in the design of buildings, not just as an add-on. It awards Flexsola the status of uniqueness among modular building systems.

The Flexsola model is illustrated in the following pages (C2-11, C2-11.1, C2-11.2). It is is one of the alternatives I was contemplating to use.

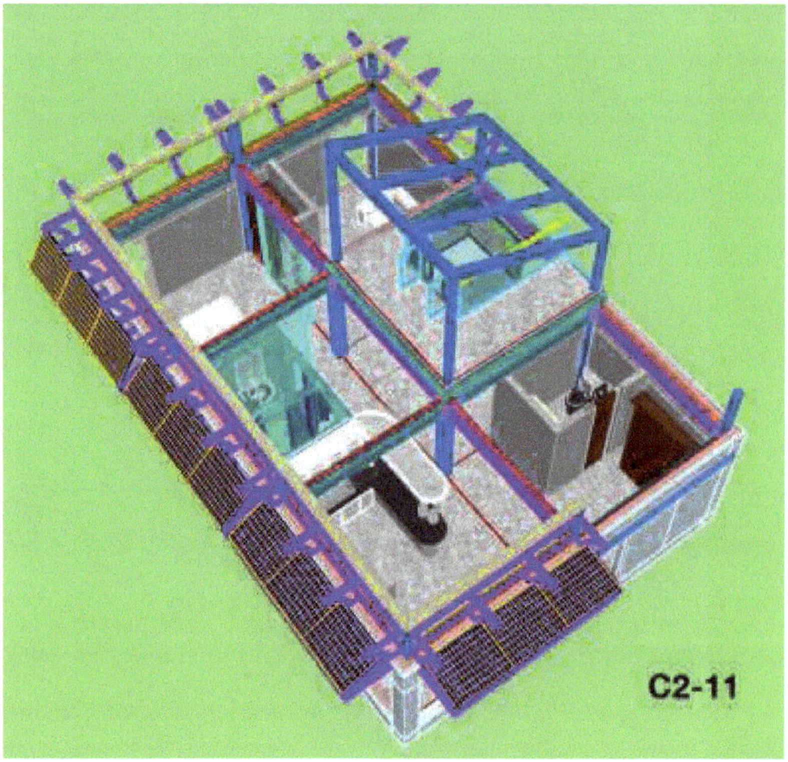


See video: https://vimeo.com/673897644

The house includes a main floor with six cuboid modules, and optional rooftop sunroom, solar awnings, DASTDs, decks, and a carport. In this case, the house is raised 0.6m from the ground to make it floodproof. The optional components can be easily added later, which makes to house more affordable than conventional homes. This concept is most advantageous in areas that are subject to flooding, which is becoming more numerous due to climate change.

In 2018, I made my first attempt to use the Flexsola modular system to build a house. I have purchased a building lot at 8326 Burwell Rd, near the famous Pinery Provincial Park at Lake Huron in Ontario, and I was planning to construct a one-storey vacation home.

Due to the flood plain condition, the ground floor had to be raised approximately 1m above the existing ground, leaving an open crawl space below. The design worked out perfectly and met all the municipal requirements.

My second attempt was to build in the beautiful Bruce Peninsula. Unfortunately, both were short-lived due to financial constraints and the Covid19 pandemic.

Following is an article in Green Building + Architecture, Volume 6, Number 4.

C2-11.1

FAST FORWARD

Flexibility of Flexsola design key to creating sustainable and environmentally-friendly housing

C2-11.2

I have also worked out the option of splitting the floor plan into segments (C2-11a & C2-11b), which are sized to fit standard containers. The segments can be preassembled in factory, transported to the site, and assembled with drones.

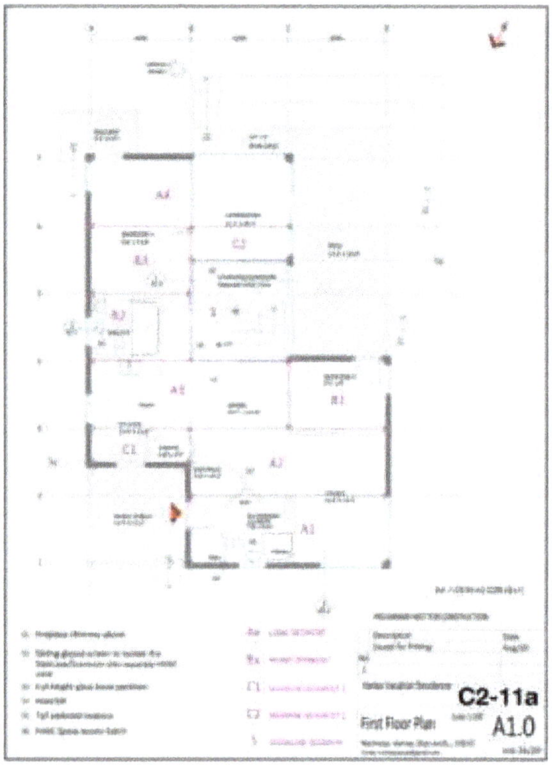


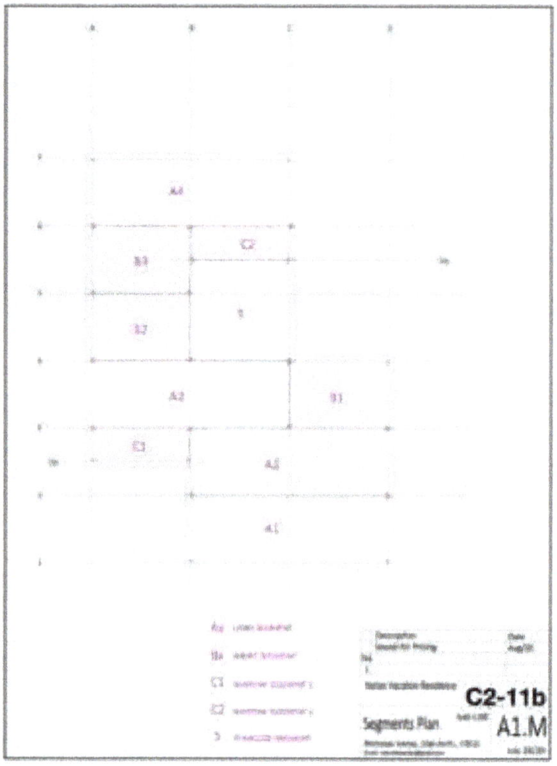


As part of my work on the concept of elevated urbanism, I designed a variation of the Flexsola House, which is raised above the ground. This could create a low-density urban community that would be adapted to frequent flooding, and which would allow wildlife to move freely, as described in Chapter 5.

In the example illustrated below (C2-12), the house is raised above the water and is accessible by boat and helicopter.

**Flexsola Housing over waterways is a disruptive concept, which could be an ideal solution for low-density residential developments in marshy areas, and near coastlines threatened by the raising ocean, such as Nederland's or coral reef islands. And why not? The construction technology already exists, and as dry land surface area recedes due to climate change, it makes sense to expand urban communities over water rather than sprawl on natural areas and farmland. The impact on marine, lakes, and marsh environments should be minimal.**

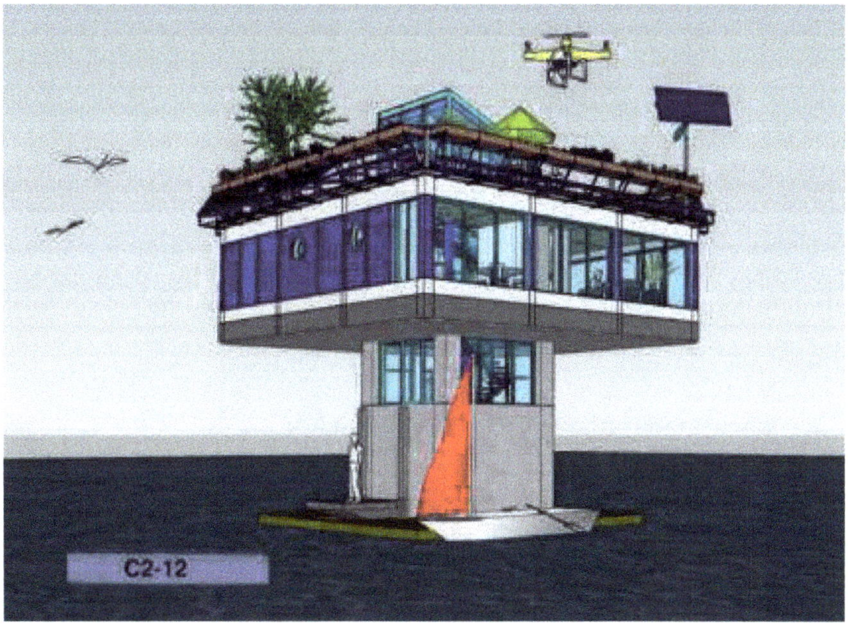


All services, which cannot be provided within each house (i.e., communication cables), can be accommodated in submerged platforms placed at a sufficient depth to avoid interfering with the boats and be made accessible to divers for maintenance. Each house shall be self-sufficient in energy, water, and sewage treatment.

In the following image (C2-13), a group of Flexsola houses illustrated above (C2-12) are accessed by an elevated street with a green rooftop protected by guardrails (not shown). The group can be expanded as desired.

See video: https://vimeo.com/697126592

# Chapter 3: Flexsola City

## The concept of Elevated Urbanism

Urbanism is the art & science of planning cities and begins with creating the big picture of a country. It is a very comprehensive process, which must include all required socio-economic and demographic data for all activity sectors, such as transportation, industry, health, education, etc. Then the view can focus on specific regions and individual cities.

**New Urbanism** is an urban design movement that promotes sustainable development. What New Urbanism and all recent planning proposals have in common is the shift of emphasis in transportation from the automobile to walking and bicycling. However, due to inertia caused by social, economic, and other factors, they continue the old paradigm of covering the natural ground.

The New Urbanism, although it brought a fresh, humane approach to urban planning, did not solve many of the problems that cities face today, and most importantly, it did not come to the rescue of the natural environment.

Imagine a world where Humankind lives in harmony with Nature! Elevated cities could help create such a world. I believe that self-sufficient urban communities built with modular building systems could elevate the human presence above the ground, thus returning the use of the land to wildlife.

Innovation in several areas of human activity, such as AI, communications, industries, and transportation, has been growing exponentially in the last few years, fueling global efforts to curb climate change. However, it has been very slow in urban planning, which continues to be frozen in the old paradigm of ground-based buildings. The main reasons are primarily historical, economic, and psychological.

As humans have evolved from apes over millions of years, they have descended from their original aerial habitat thanks to their intelligence, but they continued to depend on nature for sustenance. So, everything they built during this process, starting with their shelters, was placed on the ground or on water. Obviously, it was the most convenient, practical, and inexpensive way to use the resources readily available, earth, water, stone, and wood.

Hunting and farming have reinforced this connection with the land. The last ten thousand years have thus resulted in the paradigm of ground-based building, which turned out to be extremely destructive for the natural world and, ultimately, for humans as well. It was and continues to be the "easy" solution for fulfilling human needs.

Today, it has become evident that a major change is needed in architectural thinking and the attitude of the public to protect what is left of the natural world. Regardless of how stressful it is for some people to move from a house to a high-rise apartment, this process has proved to be inevitable and began about two centuries ago due to urbanisation and overpopulation.

Elevated urbanism is a viable solution to make the coexistence between nature and an expanding human population possible. This is not an easy endeavour, and the task is complex. Historical, economic, and psychological considerations are the main barriers.

However, the urgency of addressing climate change is a great catalyst, which hopefully will bring the necessary shift in human thinking and attitudes.

Other major barriers are spiritual and political. Changing the past building paradigm at a global level is not feasible without reconciling religions and ideologies, which divide nations and states. Here again, climate change is at work by reminding people that, as learned throughout history, survival is dependent on them working together.

Planning and designing an elevated modular city are not particularly difficult. There could be some structural engineering challenges due to the large spans required, but the building and construction technologies are already available. As a matter of fact, the rapid advancements in the fields of drone transportation and robot manufacturing and installation make elevated urbanism more and more appealing.

Modular buildings are currently gaining popularity due to their higher quality, faster delivery, and lower cost. Therefore, I have associated the concepts of modular building and elevated architecture, as I believe that the former will facilitate the success of the latter economically.

## Principles of Sustainability

The best source is the "seven generations" philosophy of the First Nations, of fulfilling the needs of current generations without compromising the needs of future generations while ensuring a balance between economic growth, environmental care, and social well-being. To implement this philosophy, there are several areas, which require special attention from urban planners and architects, and which necessitate a departure from conventional parameters.

During prehistoric times, humans were very dependent on animal and plant life and lived in a much closer relationship with nature. Some native cultures around the world have preserved this way of life up to this day. These cultures live with a balanced exchange with nature, a balance that was broken in the evolution of modern society.

Sustainability requires an equilibrium between meeting human needs and maintaining the health of natural systems. The current economic accounting system, which dominates a free market society, does not recognize the cost of pollution, reduced biodiversity, or the value of depleted resources. If it did, it would suddenly become unjustifiable.

Humankind is currently facing the dire consequences of indulging uncontrollably in the planet's natural resources. From the beginning, Man took Nature for granted, and as a result, he has been poisoning the essential elements that sustain life: the air, the water, and the earth. He has even succeeded in altering the climate and threatening his future on Earth.

Many architects, including myself, are guilty of creative arrogance at one point or another in their careers. This tendency must have its origin in the duality of architecture as art and science, and every true artist needs to be original. However, the important thing is to recognize this arrogance and control it. In Nature's eyes, the least obtrusive a human creation, the least damaging it is.

The concept of sustainable development is the touchstone for the future of the planet. In the past, scientific, and technological progress has stimulated growth. Today, embracing the sustainable dimension requires a radical repositioning of the role of research.

Towards the end of the twentieth century, the areas available for development constituted a small percentage of the total urban world.

Environmental and economic considerations have finally imposed more controls on the accelerated urban sprawl that followed the Second World War. Today, governments are facing the difficult dilemma of either abandoning the environment or stopping growth.

There is much information available about sustainability, its definitions, and its principles. The terms green, sustainable living, and, lately, regenerating buildings have been widely used and sometimes confused. Ultimately, the goal is to sustain human civilization on Earth in harmony with nature for many generations. According to the First Nations' culture, it should be for seven generations! This goal is reachable only by replacing the conventional practice of planning, design, and development with an integrated process in which the main elements involve long-term sustainability.

## The LEAN Formula

Sustainable Development must harmonize four basic elements: land, energy, architecture, and nature. Their initials make up the world LEAN, a most appropriate acronym.

The first and most important element is LAND. Land availability is limited, and therefore, planners and developers must use it wisely and efficiently as an irreplaceable commodity. Development should occur mainly in underutilized urban areas to use the existing infrastructure and services more efficiently.

Since becoming the dominant specie at the top of the food chain, humankind has assumed control over land, in direct conflict with all other creatures. It has used its superior intelligence to dominate all other animals to prosper, but in the process, it has spoiled the natural world, misusing, and wasting too much land.

For example, speculation about farms has resulted in urban sprawl, which is choking most major cities around the world. It is time this process be reversed by redeveloping intensively existing urban areas rather than continue destroying farmland and natural habitats. This must become a prerequisite for a particular development to be qualified as sustainable.

ENERGY is the second criterion for assessing the sustainability of development and should be renewable. When planning development, one must consider all the activities that generate energy consumption.

The list is long and includes extracting raw materials, manufacturing, transporting, and constructing building materials, heating, cooling, ventilating, and lighting required over the lifespan of buildings, transportation required by people, supplies, maintenance, etc. The energy evaluation of these activities is a very complex and laborious task. To be sustainable, development should be at least energy neutral; preferably, it should produce more energy than it will consume.

Thirdly, there is ARCHITECTURE, which is a harmonious marriage between art and technology in shaping the built environment. It encompasses urban planning, building design, and construction. Architectural creativity and innovation in working with land, energy, and nature are the keys to sustainable development.

There are numerous recent examples, such as the "green" roof. In this concept, the flat roof, which becomes a garden terrace supporting plant life, recreates a large portion of the land occupied by the building. A common sense and effective method for reducing greenhouse gases are to cover as many buildings as possible with green roofs. Originally initiated by a few architects, urban designers, and environmentalists in Europe, the concept has spread quickly around the globe.

Green roofs greatly contribute to reducing greenhouse gases and pollution. By substantially reducing the carbon footprint of buildings, green roofs are an important element for a balanced exchange with nature.

Finally, there is NATURE. She creates life, and as such, it must be included in any sustainable development. For example, the practice of cutting down trees to make room for buildings is barbaric and must stop. Existing natural areas must be integrated into, not replaced by, human habitat. This is possible only by adopting a new ecological urban planning philosophy. There are many examples of "green" communities and building designs worldwide which are inspired by and live in harmony with nature; they take advantage of the benefits of modern technology.

As human society has progressed technologically, it has also alienated itself from the natural world. It is a proven fact that people are happier and stay healthier when living and working in environments that simulate natural habitats.

The world should adopt the LEAN principles of sustainability: use Land wisely, consume the least Energy, and live in harmony with Nature through creative Architecture.

## Equitable Land Ownership and Rights of Use

Private ownership of the land makes sustainable planning extremely difficult, if not impossible. Most decisions take too long and do not reflect the interest of the community. Often urban planning is modified under the pressure of private interests, mainly large landowners, and developers, who seek to maximize their profits at the expense of the public.

Public ownership, on the other hand, has its own shortcomings as well. In the former communist countries, many major mistakes were made, either because of incompetence, ideological reasons, or purely just because it was too easy to make planning decisions without public input and without being questioned. Private ownership of natural resources cannot be ethically justified.

No lands, lakes, forests, minerals, or any other natural feature has ever been, or will ever be, created by Man. Yet many are privately owned today through inheritance, and their ownership originated from wars of conquest, sometimes robbery or murder. Therefore, the "title" on these lands stands on very shaky moral ground.

The means by which this ownership was obtained hundreds and thousands of years ago are, fortunately, no longer acceptable today. Most world conflicts in history have originated from disputes between different ethnic groups over land. There can be no peaceful solution to these conflicts without abandoning the traditional mentality of exclusive possession of the land.

Of course, there will be those who will argue that only private owners can properly look after the land. Unfortunately, the results of many owners' "care" are apparent in systematic destruction and neglect.

Mother Nature and the Land have been ravaged for centuries, with nobody to object except the natives. The so-called "primitive" aborigine cultures have never considered nature and land as property and have treated them with respect.

Elevated urbanism is a perfect fit for sustainable development and LEAN. But, to have a chance to succeed, governments must find an equitable balance between the ownership of land and the rights of use. Once again, the climate change crisis should be a catalyst for finding a solution, which should permit governments to approve elevated urban developments above privately owned land without the consent of the landowners, subject to certain conditions.

## Self-Sufficiency

Increased self-sufficiency is essential for achieving a sustainable society and should be a prerequisite for sustainable development and elevated urbanism. Such self-sufficiency requires a combination of renewable sources of energy. All energy sources in an elevated urban community shall be renewable: solar, wind & hydrogen, and the urban community shall be energy self-sufficient and shall not rely on the electrical grid. Fortunately, some forms of renewable energy, such as solar power, wind, and geothermal, are readily available in most areas of the world.

A solar & hydrogen system could provide a substantial portion of the energy needed for a house, depending on the amount of solar radiation. This system uses electricity to run electrolysers, which split water to make hydrogen. The hydrogen gas is then stored and used to run a fuel cell. The only exhaust is water. This is a pollution-free carbon-neutral electric system, using hydrogen as the energy storage medium. The main components are the Electricity Source, the Battery Bank, the Electrolyser, the Storage System, the Fuel Cell, and the Inverter.

The Electricity Source can be a PV Array provided by photovoltaic panels, a windmill, or a combination of both. The Inverter is a critical part of any off-the-grid electric system. It converts DC into AC current, enabling the power from solar panels, batteries, and fuel cells to run normal house loads.

Some of the surplus electricity collected from the solar panels can be used to produce hydrogen as an energy storage agent and oxygen through hydrolysis. The hydrogen can then be used in fuel cells to produce electricity during cloudy periods. The oxygen produced enriches the inside air and maintains a healthy air balance.

Despite all heat loss or gain preventive measures, including thermal insulation, any dwelling will need a backup heating and cooling system suitable for its specific geographic location. In cold climates, radiant floor heating would probably be the most efficient and comfortable way to supplement passive solar heating.

Energy self-sufficiency would eliminate the need for importing electricity from the grid; in fact, it would enable many individuals or communities to sell their surplus electricity.

Rainwater should be collected, filtered, and purified as required for each type of use: drinking and cooking, washing, and showering, flushing the toilet, and hydrolysis to produce hydrogen.

All buildings should have green roofs designed as water collectors. There is much to be learned from natural recycling processes in preserving and replenishing resources such as water and agricultural soil.

## Local Waste Treatment

Sewage should be regarded as a resource rather than a waste product: it should be recycled at the source rather than collected in large treatment plants that require high infrastructure and maintenance costs. Domestic waste can become a source of valuable organic fertilizer and hydrogen for energy. The methane by-product of septic treatment may be collected and used to produce Hydrogen.

Elevated urbanism shall replace the current practice of cumulative sewage and waste treatment plants with small, advanced treatment units with septic beds incorporated into the existing landscape. Unrecyclable waste material shall be collected in underground storage and processing rooms located near the vertical circulation towers.

## Local Food Production

Growing fruits and vegetables around the house have long been a practice in many cultures. The practice of Permaculture is gaining momentum, considering its positive impact on human health and the natural environment.

Organic food production is possible by incorporating garden plots between the elevated railway and buildings. This is the most appropriate use for this land area, where the natural landscape will be disturbed by construction and the noise from traffic.

Furthermore, in addition to playgrounds and solar panels, the rooftops could incorporate vegetable and fruit gardens. Vertical farming is possible in wheelchair-access greenhouses, which are described later.

## Sun and Light Exposure

First and foremost, the configuration of the buildings is critical for maximizing sun and natural light exposure. Continuous solid areas must be limited in size and be configured following sun studies for each project.

The use of computerized mirror systems, which can reflect sunlight as needed, can make life easier for architects, but they are expensive and should not be a substitute for planning a good distribution of natural light.

## Superior Economics and Long-Term Maintenance

- Although their initial cost is higher, elevated urban developments have a lower long-term cost than ground-based developments. The higher initial construction costs shall be recovered by savings from lower long-term maintenance and energy costs.
- The various utility pipes can be incorporated into the infrastructure of the elevated roads and railway and should be accessible for maintenance. This should eliminate the current wasteful repairs.
- The horizontal network of building ducts, pipes, and cables can be incorporated into the module ceiling space and be connected to main risers located in vertical shafts. They can all be easily accessible for maintenance.
- Heating the paving of elevated streets will become viable and will eliminate the need for snow plowing and de-icing. This will prevent deterioration and the need for concrete and R-bar repairs, resulting in long-term savings.

- Substantial cost savings can also be achieved by avoiding or delaying demolition, excavation, and cleanup of construction sites prior to proceeding with the new construction. For example, the delays caused by the uncovering of archeological artifacts during excavation will be avoided.

In August 2002, I contributed to the Mesquite Lake Eco Park Sustainable Community Conference and the Sierra Club Land and Urban Environmental Forum in San Diego. I have included my presentation material (C3-S1, C3-S2, C3-S3, C3-S4) in this book as an example of new urbanism.

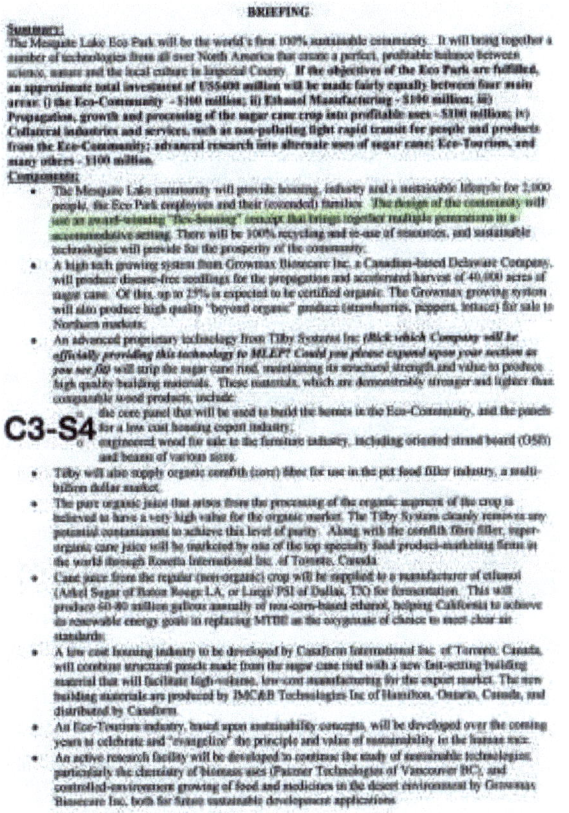
BRIEFING

C3-S4

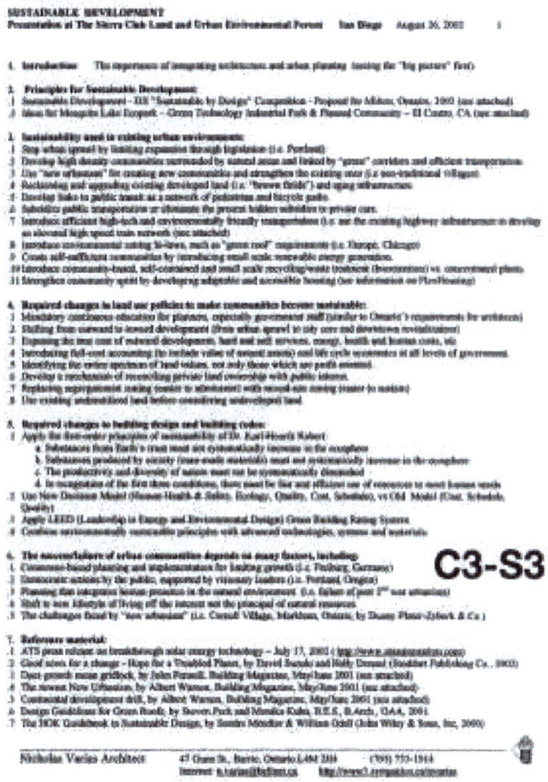
SUSTAINABLE DEVELOPMENT

C3-S3

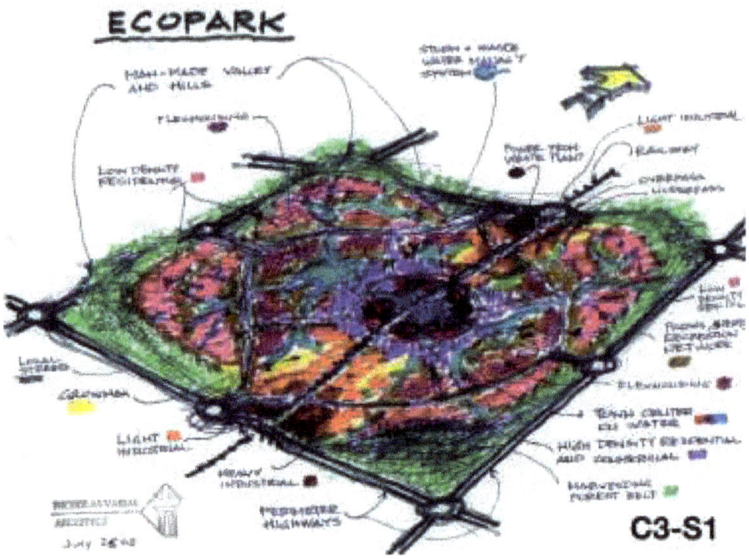

Elevated urbanism proposes to raise all new infrastructure and buildings above the ground. The anarchy (lack of proper planning) of current urban concentration, which has created the multitude of current societal problems we face today, must come to an end. Elevated urbanism is a practical way of achieving this for new urban communities and for repairing existing urban areas.

Aerial development could reclaim lands in unattractive or hostile environments where traditional housing would not be appropriate. Such lands may include infill lots near highways, industrial areas, and "brown" (contaminated) fields. Here, artificial re-grading would, in fact, rehabilitate the land and create green space. Such re-development would contribute to urban intensification, which is an essential component of sustainable development.

The goal is to create a network of urban communities connected by fast above-ground trains and highways. Each community would be developed for an optimum number of people and be served by its own business, commercial, health, education, cultural, and recreational facilities.

I believe that elevated urbanism could best achieve sustainable urban development. It can also facilitate connecting urban communities separated by highways. A good example is my 2005 proposal for the Reunification of Oakville. The Queen Elisabeth Highway (QEW), which was built in 1917, has divided the town of Oakville, Ontario, into two separate areas: the historic old town, founded in 1857, to the south along Lake Ontario, and the new urban expansion to the north. A series of over and underpasses were subsequently constructed across the QEW to serve vehicular transportation.

My concept for reunification (C3-S5, C3-S6) creates a chain of elevated shopping malls combined with offices and apartments, and developed over the existing bridge infrastructure, which would be widened. Pedestrian and bicycle traffic connecting the two separated urban areas would thus be facilitated and encouraged. It is also part of the intensified development effort, which contributes to sustainability.

## Advocating Trains over Highways

I wrote and drew the sketches for the following text more than ten years ago, and I still believe my idea could greatly improve transportation.

When arteriosclerosis is left unchecked, heart failure is inevitable. The transportation systems constitute the arteries and veins of the urban body. Their clogging undermines the health of urban life and prevents sustainable development.

A good example is the increasing volume of traffic through Ontario's "golden horseshoe," which is an international artery. Here the exchange of goods between Canada and the USA and the growing number of visitors to Niagara Falls are compounded by the high number of commuters in the grater Toronto area. The main problem in "clearing this artery" is the environmental impact of expanding the existing highway or of developing a parallel railway.

C3-01

Traffic demand on highways could be greatly reduced by incorporating an elevated Light Rapid Transit (LRT) monorail over the median and/or shoulders (C3-01). This would not require additional land, and the environmental impact would be minimal. The LRT would use the highway infrastructure and the land it occupies more efficiently. By prefabricating the entire structure supporting the LRT, its erection would cause minimal disruption to highway traffic.

A specially developed crane, completed with a drilling machine, could access the median from the shoulder without having to close any lanes. The system of piers and beams could support a conventional electric train, yet it would make sense to provide for possible future conversion to the higher speed and safer magnetic levitation system (MAGLEV).

The concept has other advantages as well, such as sheltered roads that do not need ploughing and salting in winter and safer wet driving conditions. In addition, a solar collection sail installed above the railways could generate electricity and provide lighting to the highway below. The raised infrastructure could also support high-tension wires to transport electricity and communication lines.

Firstly, it should be the role of governments to invest in the research and development of such concepts. Secondly, it would make sense to invest in manufacturing facilities that would produce all the necessary components. A partnership between governments, major contractors, precast manufacturers, and steel fabricators could develop these in a cost-effective way.

The initial high cost of the factories and equipment would be a one-time long-term investment. Considering all the elements involved, such projects could bring a much needed and long-term economic boost. Finally, Mother Nature would benefit.

The most harmful effect the highway network has caused, even more than noise and air pollution, is the fact that it has fragmented natural areas. The world's flora and fauna, originally functioning as cohesive living organisms, have been split up into ever-shrinking islands with growing physical restrictions on wildlife.

In time, transferring the traffic load from the ground to the elevated trains and maintaining only two-lane low-speed roads for local traffic, service, and access to train stations and parking lots, would scale down the old highways and correct the damage done to the natural environment. Wildlife will be free to cross and move freely again throughout the land.

As illustrated next (C3-02), Mobility Interchange Hubs (MIH), built above the highways at present carpooling parking lots and overpasses, would connect all modes of transportation (trains, streetcars, buses, automobiles, motorcycles, bicycles, and pedestrians). Each MIH could incorporate a mini mall with essential services, public amenities, and commercial space. This would open the door to the creation of a local Public-Private Partnership for its development.

C3-02

Thanks to its minimal environmental impact, such concepts are an ideal solution for global sustainable transportation.

There are numerous elevated light trains in the world, already running or in the planning stage, yet this concept of redeveloping the existing highway infrastructure is uniquely disruptive, which is exactly what the world currently needs. An elevated railway would make rapid public transit along the highways more convenient and less costly than driving; this is the most effective way to break the vicious circle of unhealthy, polluting, and inefficient transportation that has trapped commuters worldwide since the expansion of the private automobile.

Such a revolutionary transportation system could help people abandon their dependency on automobiles. The present measures taken by governments are insufficient because bad habits seldom change voluntarily. Higher gasoline prices discourage driving, but for most people, there is no practical alternative to driving to work. It should be the role of smart governments to first create a better alternative. What wiser choice of spending could there be than for a better and healthier quality of life? It will undoubtedly bring great economic, social, and health dividends. Let the metamorphosis of highways begin!

## The Flexsola City idea

The Flexsola™ modular system can be applied to large buildings, specifically to large, elevated buildings. The principle of the saddle connection works the same way, regardless of size. Its advantages of flexibility and easy installation are even more evident in elevated buildings, where large spans are desired.

Following is my initial hand sketch (C2-03) of an imaginary Flexsola City, illustrating how most of the natural landscape could be preserved. Wildlife could roam freely over the land, undisturbed by buildings and roads.

I could only imagine how much this could achieve in rehabilitating natural habitats and saving endangered species. It would be like a new lease on life for Mother Nature!

Far from being an impediment, the shade cast on the ground by the elevated infrastructure and buildings would provide shelter from excessive sun radiation to many species.

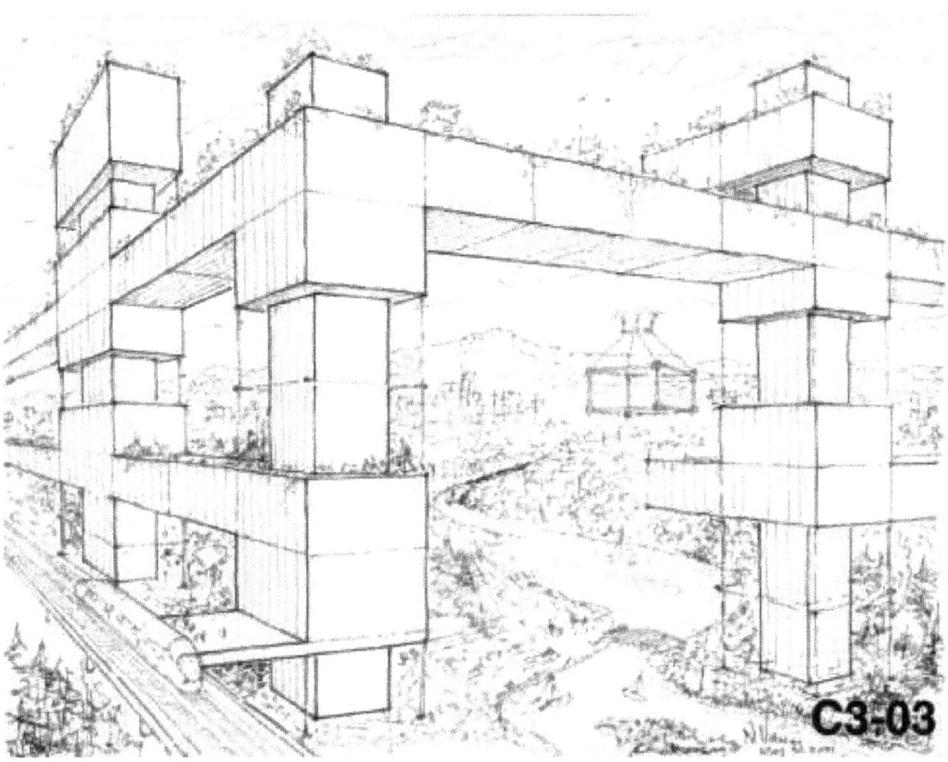

## Structural Considerations

The main objective is to reduce the number of support towers for the purpose of reducing the footprint of elevated streets and buildings and minimizing the impact on the existing landscape. This poses geotechnical and structural challenges for the engineers and is very much dependent on local soil conditions and earthquake zoning. For example, for building on deep permafrost, innovative foundation methods are available, such as rafts or floating slabs.

To achieve the Flexsola™ flexibility, only bolted demountable connections shall be used. The concept relies on large-span rectangular space trusses that apply the Flexsola™ saddle connection system (C3-04).

The structural design shall consider the performance of the entire building as an assembly of space modules. This results in large spans, which can be achieved to support elevated streets.

As stated earlier, I need to emphasize that reducing the number of supporting towers is critical for the success of the concept, as it lowers the overall footprint of the development.

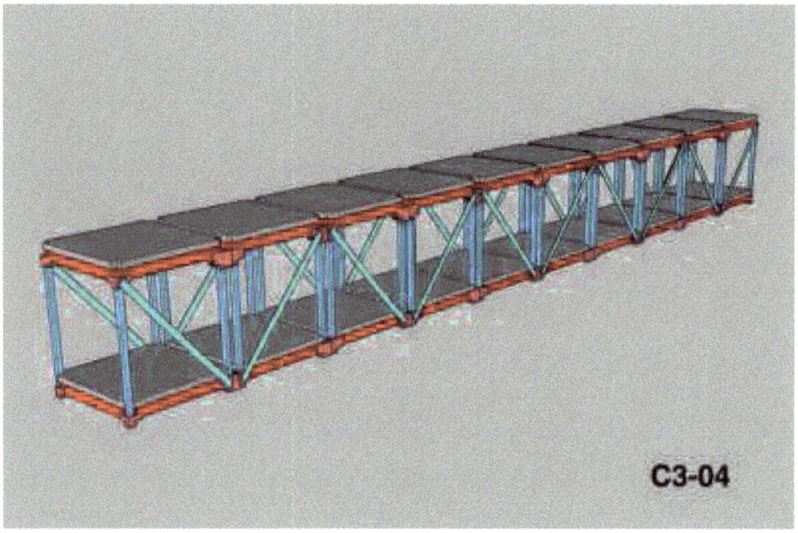


The image above (C3-04) shows are space truss constructed with Flexsola™ modules, which are engineered for the spans and loads required. The modules are assembled on the ground and then lifted and connected into a rigid space structure by drones. As described in Chapter 1, the patented saddle connection provides an ideal bolts-only connection system, which provides flexibility, speedy installation, and cost-effective construction.

The Flexsola™ modules could be assembled with robotic drones (C3-05).

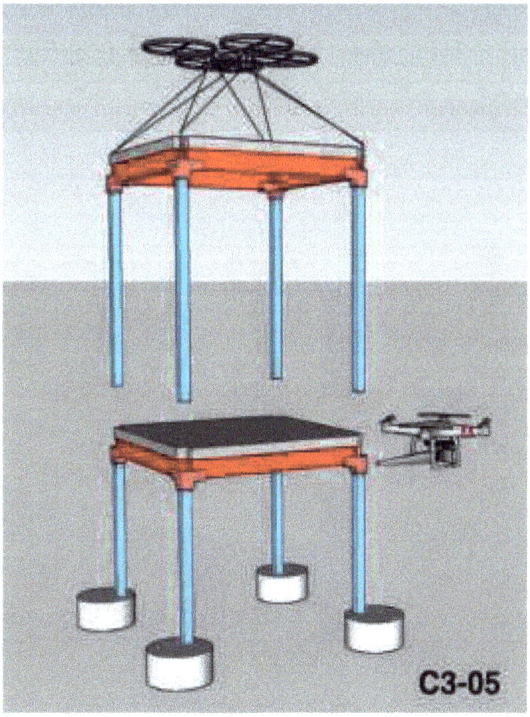

C3-05

The cuboid module allows for unlimited building configurations. An elevated pyramid is just an example (C3-06).

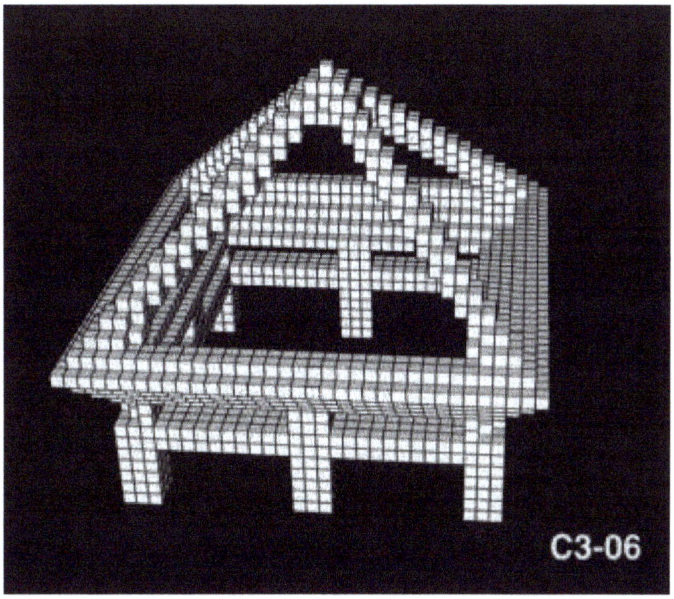

C3-06

For preserving natural areas and farmland, the Flexsola City idea could be used for developments in the desert, in flood-prone areas, reclaiming underutilized, dilapidated, or contaminated lands, and for developments over disaster areas, where construction could proceed before demolition and cleanup.

The following image (C3-07) illustrates a hypothetical city under construction in the desert.

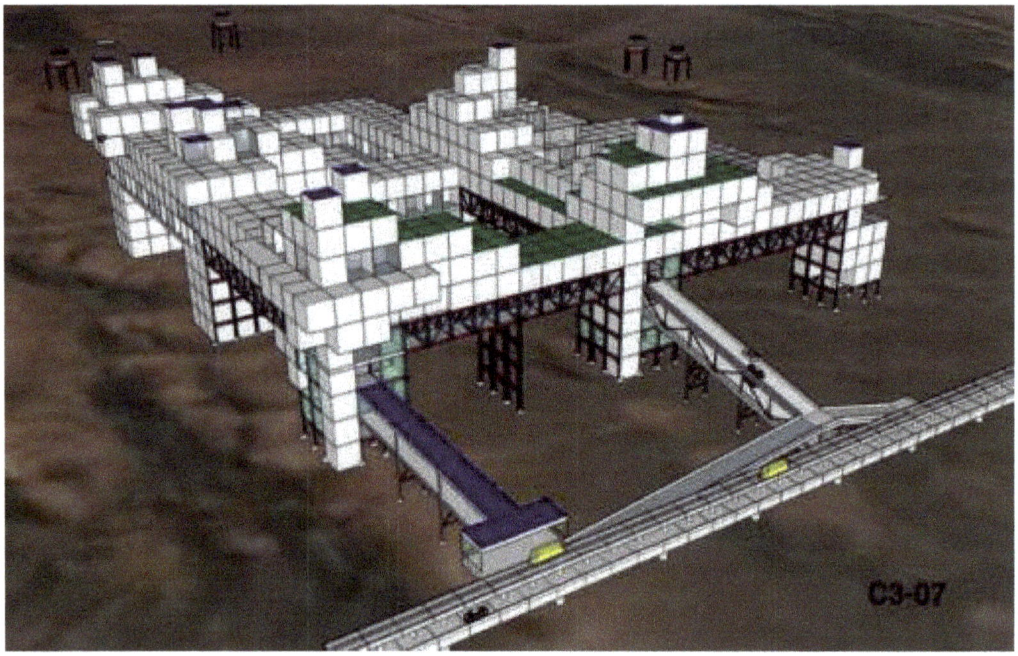


See video: https://vimeo.com/697094380

## Planning and designing Flexsola City

The entire urban development is elevated above the natural grade for minimum impact on the natural landscape.

The configuration of the buildings must account for allowing sunlight to reach the ground as much as possible. Therefore, continuous solid areas must be limited in size and be established through a sun study for the project location.

The area between the elevated rail and buildings could incorporate garden plots for the residents. The rooftops could incorporate playgrounds, gardens, and green roofs. This will result in a net+ landscaped area (after subtracting the supporting towers).

The basement of each support tower could accommodate utility and service rooms, stairs, and elevators. The ground level could incorporate a lobby, garden amenities, stairs & elevators, and service, waste & recycling rooms.

The support tower's upper level has the main lobby with 2 stairs & elevators. The entrance has a car drop-off from the interior street, which is connected to the elevated road. The lobby also provides access to the pedestrian bridge leading to the elevated light rail station. The 7m (grid dimension) cube modules can accommodate commercial space with a mezzanine, 2 stacked apartments, or two-level suites.

## Future construction methods

The high efficiency and low cost of the Flexsola™ modular system relies on the use of drones for installation. The manufacturing of structural components with 3D printing using metal powder from recycled steel shall ensure the environmental superiority of the Flexsola™ modular system.

## Elevated transportation

The elevated streets providing access to the tower's upper lobbies shall be connected to existing surface roads, some of which could be gradually decommissioned and returned to nature.

The elevated light rail shall connect the regional transportation hubs with other elevated urban communities. Tunnels and/or bridges may be used in certain conditions, such as mountainous areas.

The interior streets shall be mainly for pedestrians and bicycles, with vehicular traffic limited to residents, driverless cars, and emergency vehicles. In addition to providing access to entrances to the residential and business blocks, the interior streets could also serve as 1 and 2-storey commercial suites and shall have parking and loading spaces. The paved areas shall incorporate heating cables to melt the snow and ice.

## Energy

The urban community shall be energy self-sufficient from renewable sources, such as solar, wind, and/or hydrogen, and shall not rely on the electrical grid. This self-sufficiency is facilitated by incorporating solar PV systems into the design of the infrastructure and buildings.

The infrastructure of the elevated streets can accommodate a multitude of energy-producing systems, such as PV inverters, batteries, hydrogen storage, etc., all easily accessible from below without disturbing the pavement.

## Services

All service pipes can be incorporated into the ceiling space of the structure supporting the elevated roads and light rails and are easily accessible for maintenance from below without disrupting the traffic. This gives elevated infrastructure a great advantage over the conventional, and I would say the primitive practice of burying the pipes under the pavement. It is time that infrastructure catches up with the levels achieved in other sectors of the economy, such as communications.

## Waste Treatment

Small sewage treatment units with septic beds incorporated into the existing landscape are located near the towers supporting the buildings. Unrecyclable waste material is stored underground in the support towers and is transported through tunnels to a processing plant. The objective is to recycle 100% of all waste produced in a self-contained community.

### Economic benefits

The higher initial construction cost of elevated urbanism shall be recovered from lower long-term maintenance and energy costs.

There are many advantages resulting from raising the infrastructure and buildings above the ground, such as:

- Heated paving can eliminate the need for snow plowing and de-icing, which will substantially reduce deterioration and the need for concrete and R-bar repairs.
- The paving can incorporate PV panels integrated into the community energy system.
- The plenum space of the street infrastructure permits convenient and low-cost maintenance of the paving substructure, service pipes, ducts, energy-producing systems, etc. Compare this with the current primitive practice of road and highway maintenance, which results in traffic nightmares and pollution, and escalating costs.
- Practically unlimited flexibility in expanding or modifying service and energy systems.
- Avoiding or delaying demolition, excavation, and cleanup of construction sites prior to proceeding with the new construction.
- Reduced delays due to archeological artifacts uncovered during excavation, when in conventional development, the work comes to a halt until release is approved by archeologists.

**Elevated Urbanism, exemplified in this book by Flexsola City (Chapter 3) and Honeycomb City (Chapter 5), is a disruptive concept that I believe is necessary for ending the fragmentation and destruction of the remaining global natural environment.**

# Chapter 4: The Honeycomb

## Background

Since childhood, I have been fascinated by the world of bees and especially by how they construct their homes. It appears to be a miracle of cooperation between nature and geometry. The hexagonal structure of the honeycomb is one of the few examples where nature is using mathematics to create extraordinarily efficient habitats.

Although natural or domesticated bee honeycombs are geometrically different from a geodesic dome, the structural strength and material use efficiency are similarly great due to the properties of the equilateral triangle component. Following is a quotation from the Buckminster Fuller Institute:

"One of the ways Buckminster Fuller ("Bucky") would describe the differences in strength between a rectangle and a triangle would be to apply pressure to both structures. The rectangle would fold up and be unstable, but the triangle withstands the pressure and is much more rigid--in fact, the triangle is twice as strong."

It makes, therefore, common sense to affirm that the hexagonal grid is superior to the rectangular grid, which has dominated the past ten thousand years of building, and that the time has come to recognize this fact while the search is on for more environmentally friendly designs for the built environment.

My interest as an architect in the hexagonal grid began over twenty years ago when I designed the Am Shalom Synagogue in Barrie, Ontario, with all walls and partitions running at 30° & 60° angles (C4-01, C4-02).

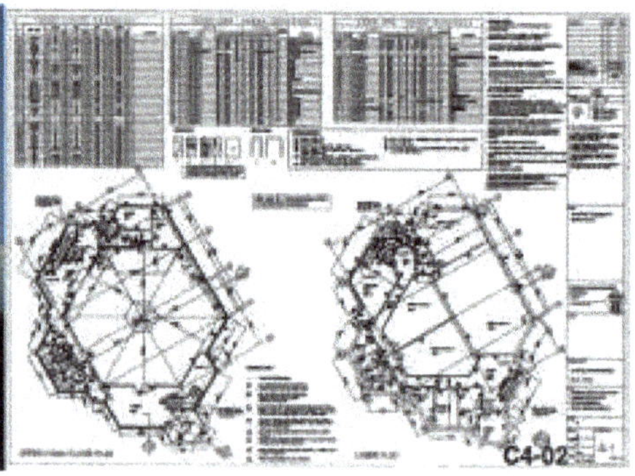

Then came Honeycomb Villages, a concept for an underground development in the desert, for which I proposed hexagonal mushroom-shaped elements with a landscaped top.

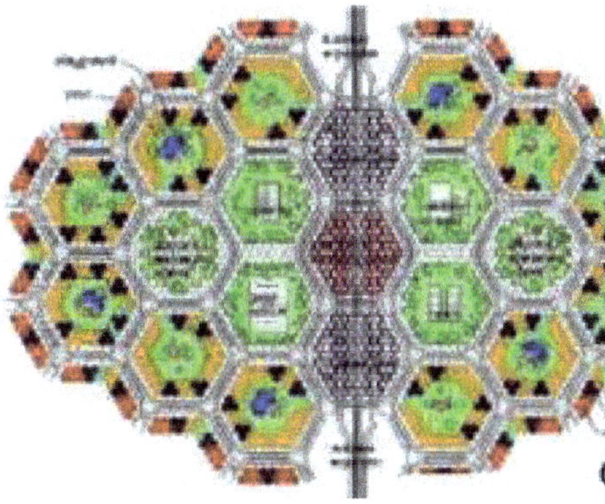

C4-03

C4-04

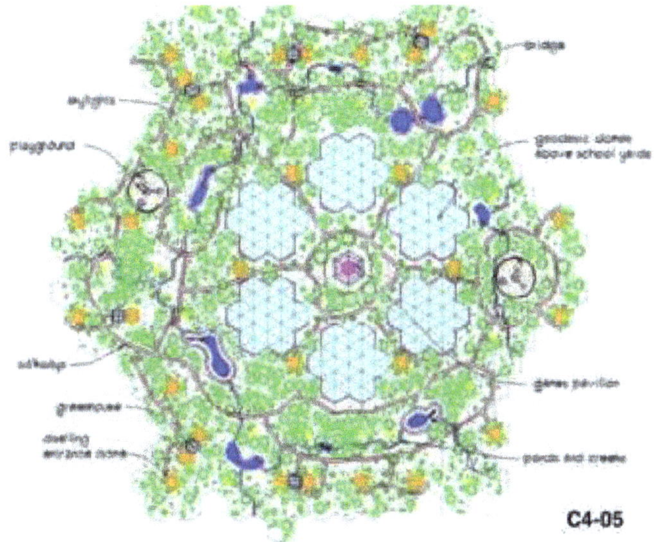

C4-05

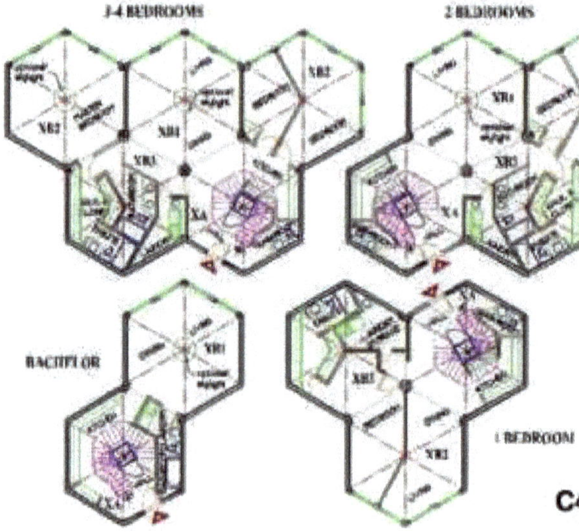


C4-06

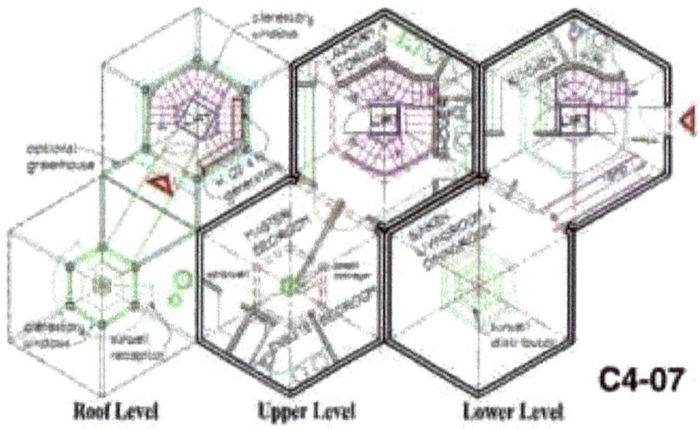


C4-07

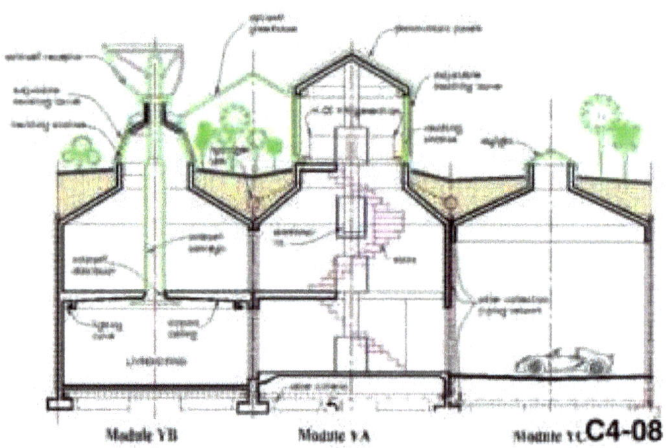


C4-08

A couple of years later, the same mushroom element became the central piece for my proposal to rescue villages on coral reef islands (C4-09, C4-10), threatened by the rising sea level. I became aware of this problem during my visit to Panama in 2008 when I met with my architect friend Howard Walker, who retired there with his Kuna Indian wife.

## Design Criteria

The modular designs I present in this book are based on the following criteria:

**a)** The number of components used to create a building module shall be kept to a minimum to reduce manufacturing, transportation, and installation costs. This is a fundamental objective for modular buildings, which I have used for designing the transportable module option for the Flexsola House and the single-family module for the Honeycomb House.

**b)** For commercial developments, the modules must be broken down into smaller components with a maximum width of 2.40m (8ft), which could be transported by ship, train, or road and assembled on site. This somewhat negatively affects the economics of prefabrication and installation. However, keeping the modules small allows more building design flexibility.

**c)** All components shall be transportable by truck, train, and/or ship, shall be assembled on the construction site into modules, and then lifted and set in place with minimal manual handling required. The lifting and installation of the pre-assembled modules could be performed with conventional methods, such as cranes, or with advanced methods. Robotic drones could be developed to install or remove bolts without manual intervention. This would greatly speed up construction and reduce costs.

**d)** All components and all modules shall be installed with bolted connections, without welding, to provide demountability, reusability, and/or replaceability as required over the life of the building. This objective requires more design effort and structural engineering as compared to conventional welding. However, the benefits are worth it.

**e)** The materials used for the manufacturing of all components shall be durable, non-combustible, and shall not produce or facilitate allergens.

**f)** The structural design shall consider the performance of the entire building as an assembly of space modules. This objective is critical for achieving big spans without large structural beams.

**g)** The modules shall accommodate all building services, such as HVAC, plumbing, electrical, communications, etc.

**h)** All components shall be long-lasting and protected against corrosion.

**i)** The composition and configuration of all components shall facilitate 3D printing, as new technologies are being developed to metal powder to replace steel for the manufacturing of structural components.

I have started my design research with a small hexagonal module with 1142mm sides, appropriate for housing, which could be easily transported. However, using such small modules to incorporate a kitchen or a bathroom proved to be a real challenge.

**Honeycomb Modules 2900**

The Honeycomb module 2900 (C4-11) is based on Canadian patent application nr. 3,145,653. The invention consists of a modular system, which creates one type of triangular module to assemble hexagonal building modules of various sizes, and which can be added, removed, or relocated within a three-dimensional hexagonal grid by using only bolts.

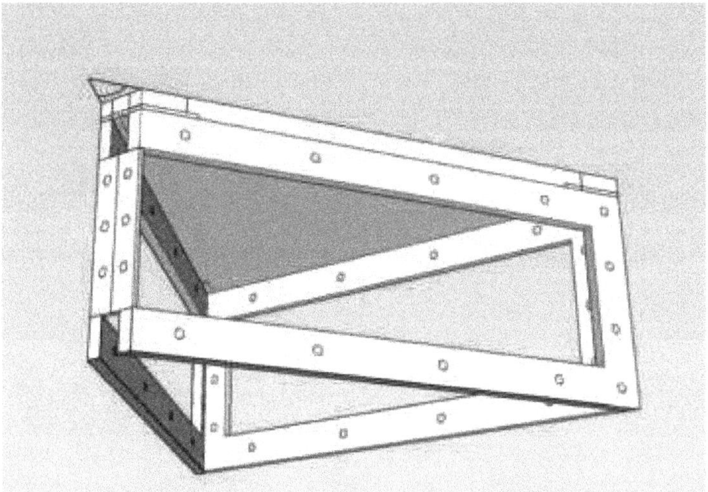

C4-11

The 2900 triangular module is shaped as an equilateral triangular prism with one chamfered corner and fits into an equilateral triangular grid with 2900mm sides. It is prefabricated from a one-floor panel shaped as an equilateral triangular prism with one chamfered corner and a metal framing assembly consisting of three vertical metal frames with holes for bolts. The floor panel incorporates four horizontal metal plates placed at its bottom corner, and the spaces above the metal plates are filled with four removable fillers.

Two of the vertical metal frames converge towards the chamfered corner and have vertical metal plates with holes for bolts. All vertical metal frames are welded to form a rigid assembly, and the horizontal metal plates have holes for bolting to the vertical metal frames. These holes also serve for attaching cables for lifting floor panels, triangular building modules, or hexagonal building modules.

The module is shaped as an equilateral triangle bases prism with 2.90m (9'-6") sides and 4.0m (13'-1 ½") height. This module, which has three equal sides of 2900mm, can be used to create hexagonal modules with 2900mm sides.

The 2900 hexagonal modules can be used to create larger hexagonal modules with 10040mm sides. The structural members of each triangular module are factory welded together into a rigid assembly.

The modules are supported by a column assembly consisting of a round column, a coupling, and a horizontal hexagonal plate. Six triangular modules are connected through a node to form a hexagonal module, which can telescope on top of the column assembly. All modules have an open-web structural flaming, which provides flexibility in the horizontal distribution of building services. Furthermore, any triangular module can be removed to create a vertical shaft for the vertical distribution of building services. This invention creates excellent design versatility, facilitates the prefabrication of buildings for efficient and economical transportation, and achieves fast assembly or disassembly of building modules.

The triangular floor slabs are bolted to the triangular metal framing to permit their removal as required later to create vertical service shafts. The tridimensional space framing formed by the joined modules is accessible from below and provides great flexibility for horizontal service ducts, pipes, and cables.

After being shipped to the building site in standard flat-rack containers, the triangular modules are bolted together on platforms placed on the ground to form larger hexagonal modules. These are then lifted and set in place to assemble the entire building structure.

The triangular and hexagonal modules allow the expansion of buildings in three directions, following a 60° grid and vertically, by using column extension couplings.

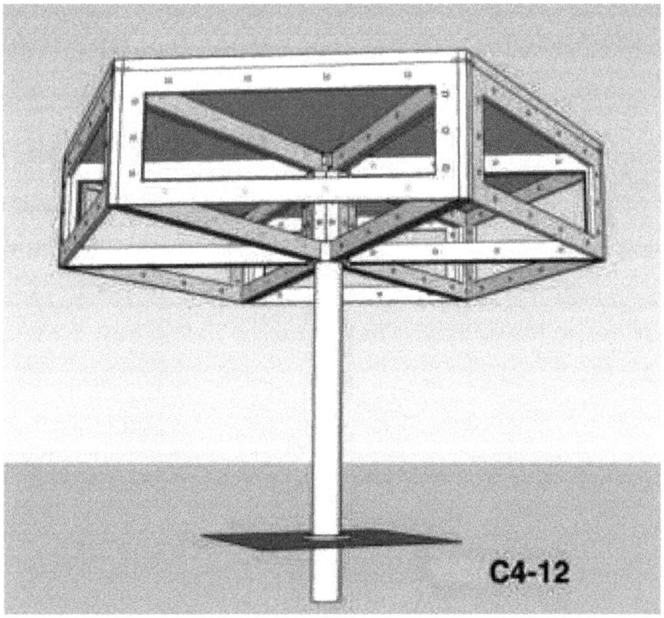

The 2900 hexagonal module (C4-12, C4-12a) is composed of six 2900 triangular modules, a metal node, and a metal column assembly. The metal column assembly includes a round metal column, a metal coupling, and a horizontal metal hexagonal plate. The metal coupling is shaped as a hexagonal prism with a cylindrical core, which slides on top of it and is welded to the round metal column.

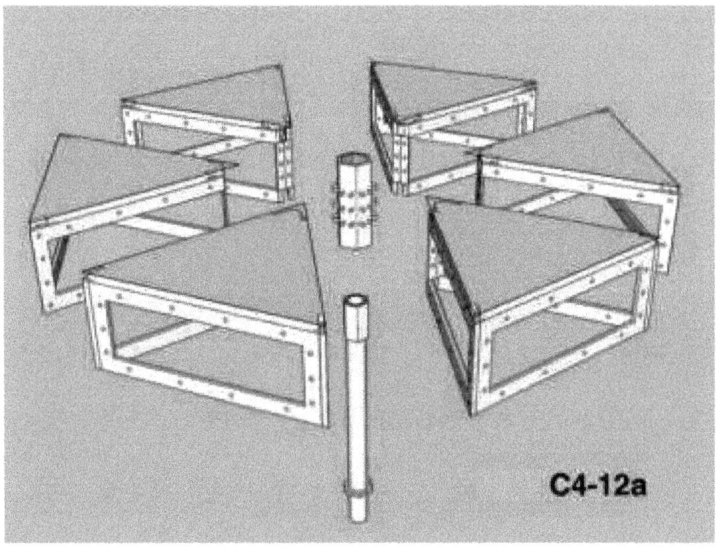

The 10046 hexagonal module (C4-13, C4-14) is assembled from three 2900 hexagonal modules and six 2900 triangular modules.

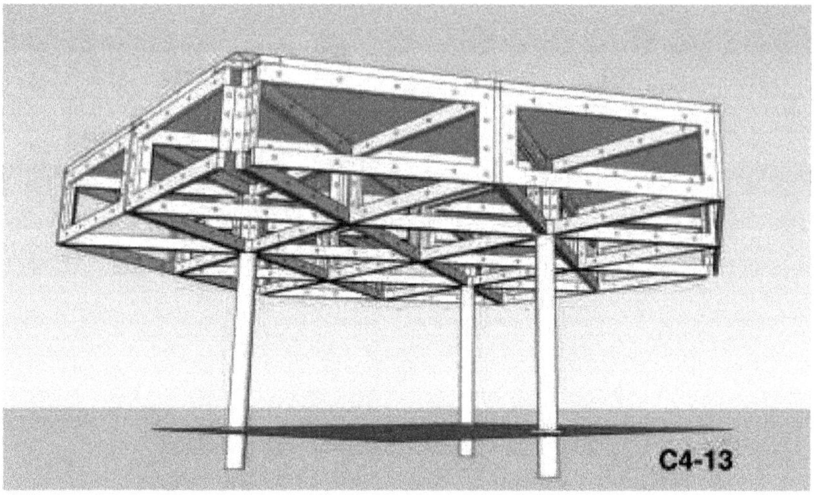


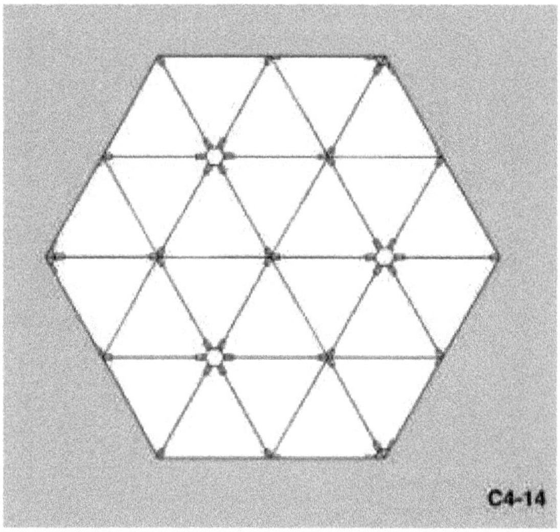


Modules can be easily added for horizontal and/or vertical expansion (C4-15).

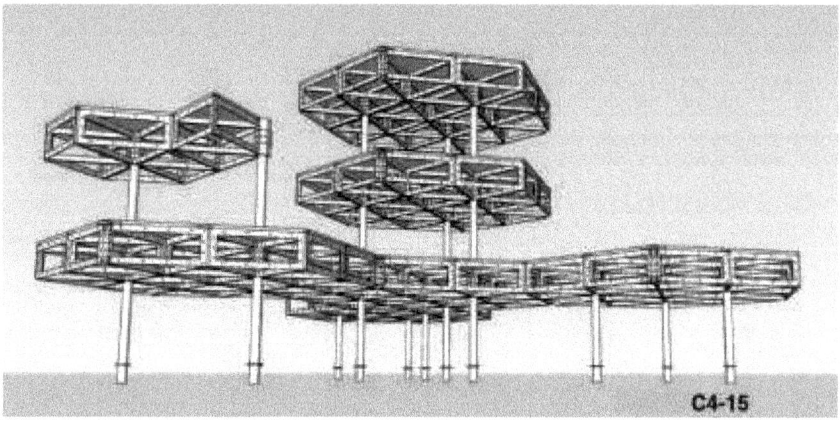

As illustrated below (C4-16), the service ducts can be installed through the space framing of hexagonal modules, and vertical service ducts cross through the space created by removing some of the 2900 triangular modules.

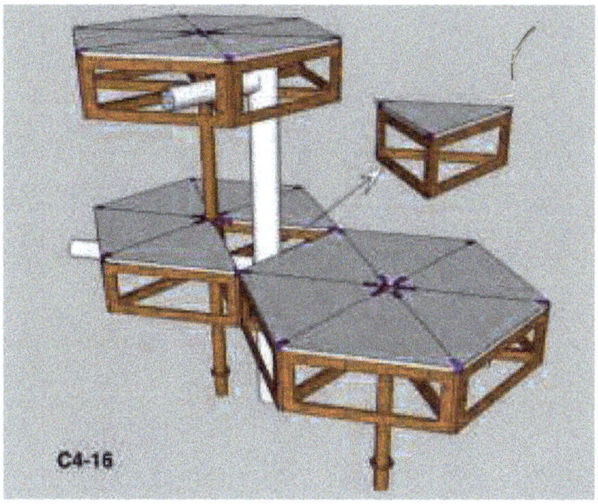

## Using The Modules in Design

Designing with a hexagonal grid is not easy. Most architects have been conditioned to think "square." Therefore, trying to accommodate all the functions, the 60° angle could be a challenge, at least at the beginning. As switching to a completely different geometry is combined with elevating the building from the ground, the task becomes even more complex. However, after "breaking the ice" formed by years of training and experience, escaping from the box becomes easy and rewarding.

## The Honeycomb House

Elevating the house above the ground adds to the vertical circulation but achieves many benefits. A growing percentage of the world population is already prepared to live above the ground and accepts stairs and elevators. One of the main benefits of an elevated building is reducing its physical footprint on the ground and thus minimizing its impact on the existing landscape. The Honeycomb House (C4-17) I present here is built with hexagonal space modules transportable by standard flat rack containers and has a 225m2 (2422 sq. ft.) gross floor area, including a rooftop greenhouse and staircase landings with accessory areas. See video: https://vimeo.com/697134607

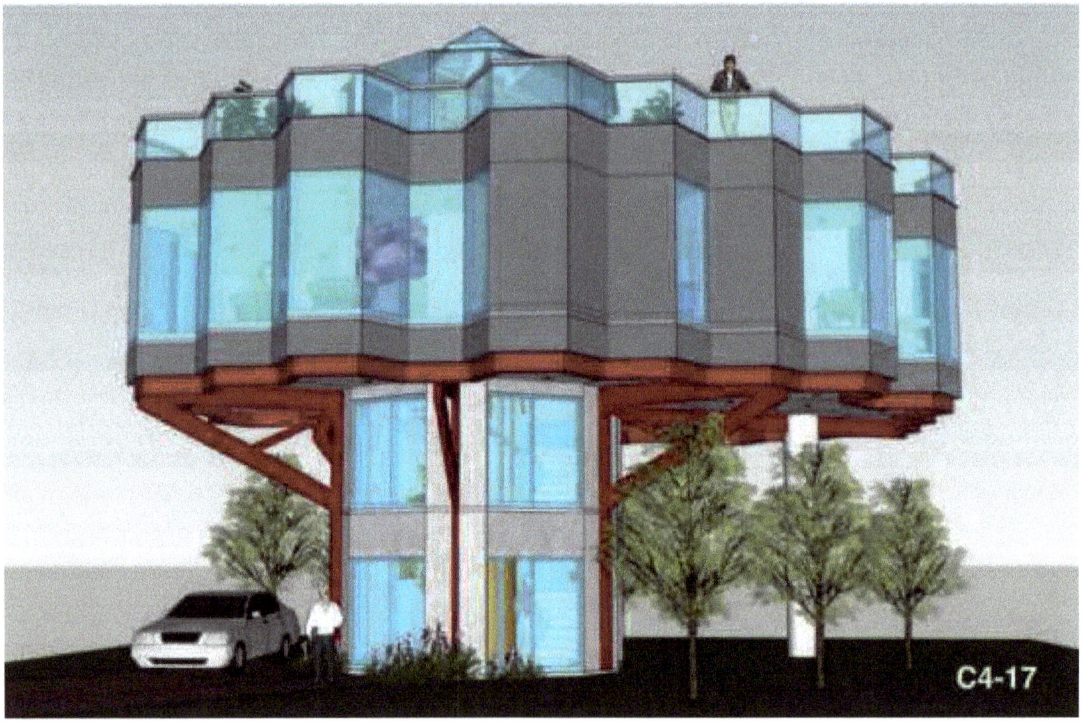


Being an architect, not an engineer, I have made certain structural assumptions based on my education and experience, assumptions that will have to be confirmed by "Limit State Design" calculations. The main structure includes an exposed steel framing supported by the concrete walls of the central staircase.

The steel structure supports the assembly of cantilevered hexagonal space modules with 1142mm sides. The modules have a concrete slab bearing on metal space framing, and if needed, hexagonal posts can also be incorporated.

The Honeycomb House, just as the Flexsola House described in Chapter 2, facilitates the adoption of an environmental and healthy lifestyle.

This includes rooftop gardening (C4-18). It can also be developed on flood plains and water, with access by boat or helicopter. The walls around the entrance floor located at the water level, which fluctuates with the weather, are waterproofed and designed to resist the water pressure.

Instead of the driveway adjacent to the entrance, the house would have an adjustable dock for boats.

The hexagonal grid creates a spectacular three-storey high interior courtyard, visible from all levels. There are no right-angled corners, and as shown above, all rooms can be well furnished (C4-18a).

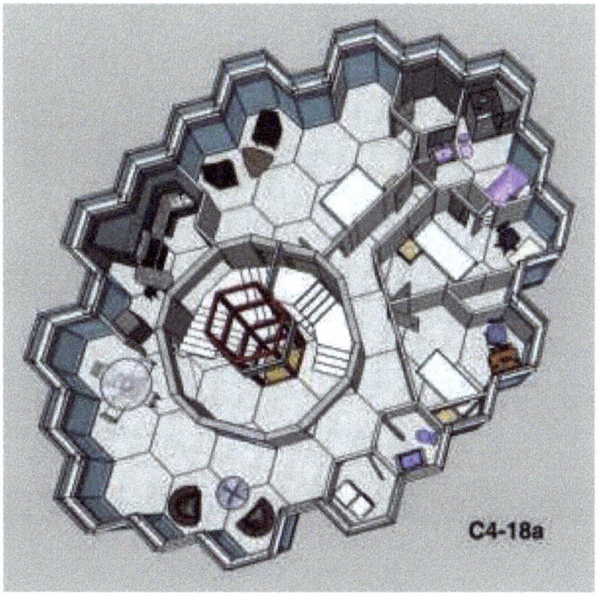


There is a master bedroom, two bedrooms, a kitchen, living, dining, ensuite, powder room, and laundry. All these rooms have windows and are accessed from a circular space around the courtyard. The central steel structure, which forms a hexagonal prism, can accommodate an optional round elevator, depicted in orange colour.

The modules are packaged for protection during transportation (C4-21) and for stacking on flat-rack containers. Then, they are lifted, set in place, and installed with drones (C4-19, C4-20).

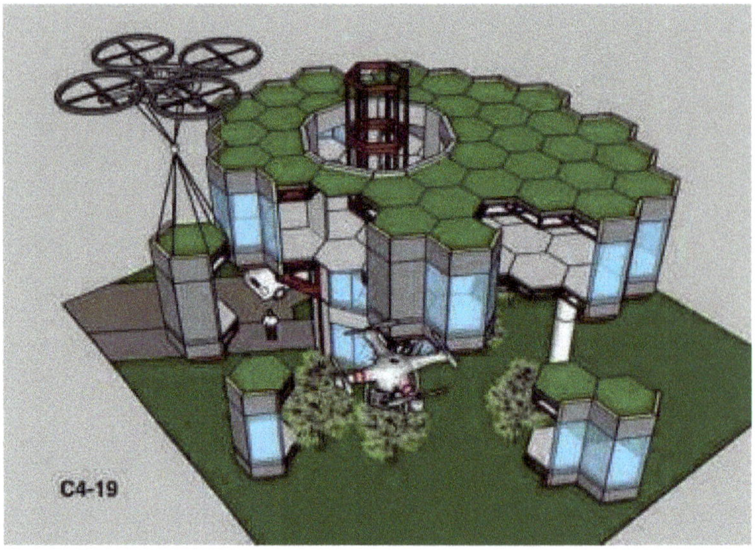

C4-19

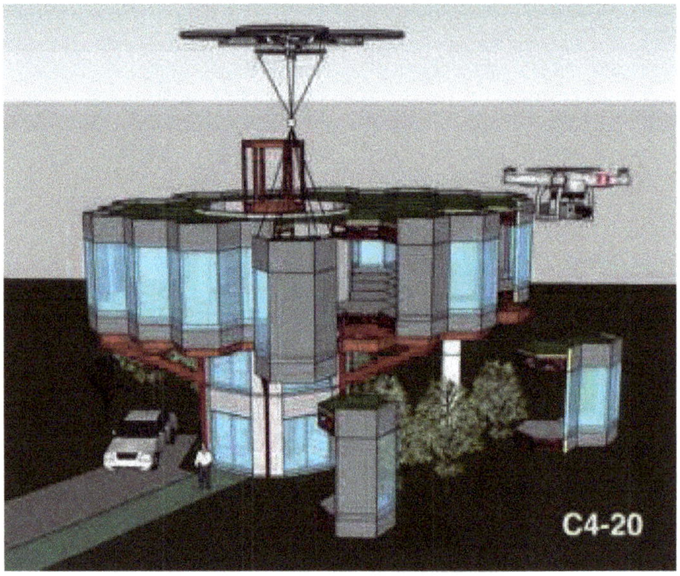

C4-20

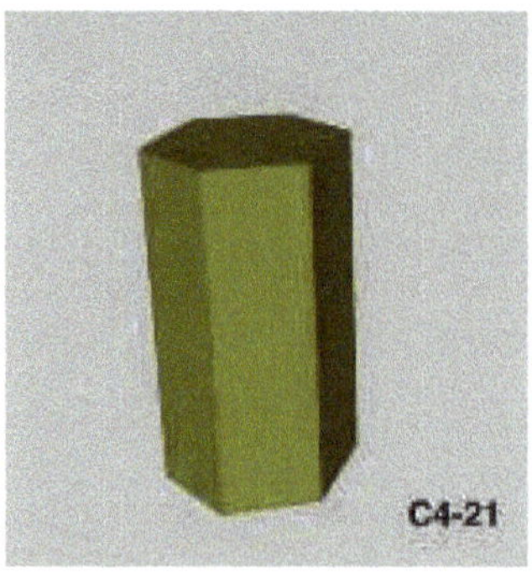

C4-21

The following images show the sequence of installation on site, which can be performed without having to cut the existing trees.

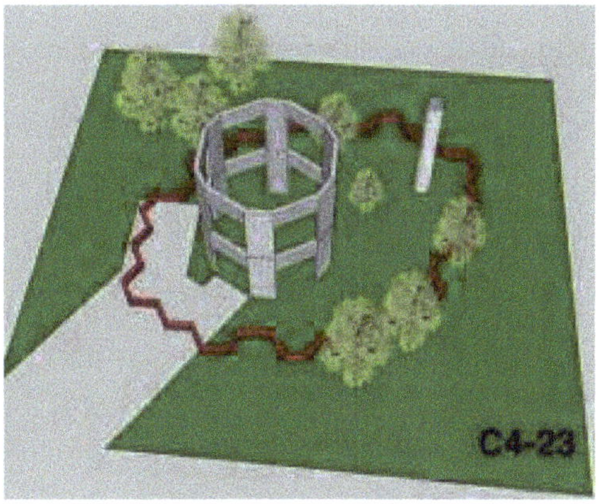


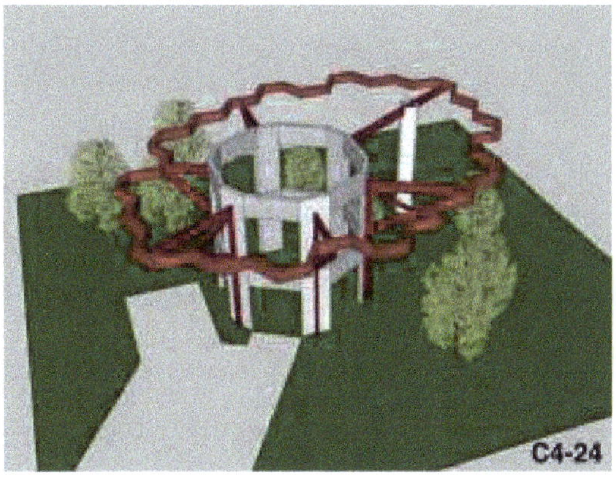


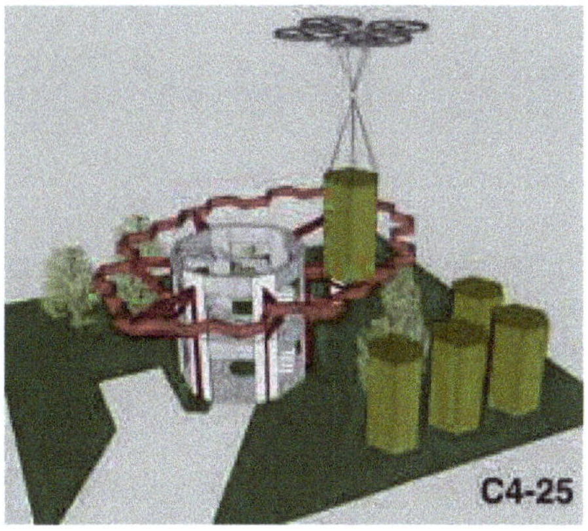

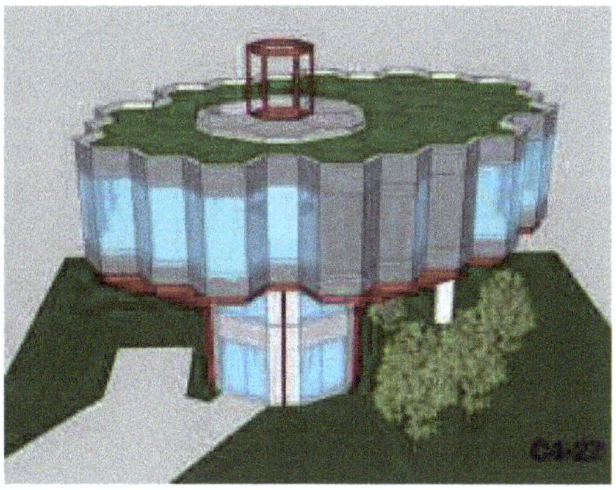

# Chapter 5: Honeycomb City

The second half of the 21st century has seen the advent of several progressive urbanistic trends, which have all claimed to achieve sustainable development. There are many definitions of "sustainable," and they have been used and misused by real estate developers to get the green light and support for their projects. However, are they truly sustainable? Do they address all required urbanistic criteria? My answer is NO, because they all focus primarily on the short-term needs of humans, and they miss the importance of the natural world. What is missing from the equation is the "inconvenient truth" that the natural world is essential for long-term human survival on Earth.

Up to this date, I could not find a cost estimation of the destruction inflicted on our planet by the global human society over the last few hundred years of modern history.

## A Magical Geometry

The hexagonal grid is most appropriate for creating an aerial habitat, as it is geometrically similar to the honeycomb structure produced naturally by bee workers. It is more flexible than the orthogonal grid, as it allows expansion in three horizontal directions rather than the two directions allowed by the orthogonal grid. This is a major advantage that outweighs the drawbacks caused by its more complex configuration, which requires additional design and manufacturing efforts.

The hexagon is a very versatile geometric shape that can create fascinating images. This ability facilitates the design of several alternatives to module configuration and connection details. Each alternative ought to satisfy the specific requirements of use, fabrication, or installation method and the structural solution for the overall building.

## Objectives of Elevated Urbanism

There are several specific targets for elevated developments, such as protecting nature, preserving farmland, reclaiming the desert, recuperating misused lands, and developing overwater or flood-prone areas. The feasibility of such developments depends on resolving first the legal implications of land ownership for each project.

This chapter presents only a few scenarios of elevated honeycomb communities, but the design configuration options for architects are practically limitless.

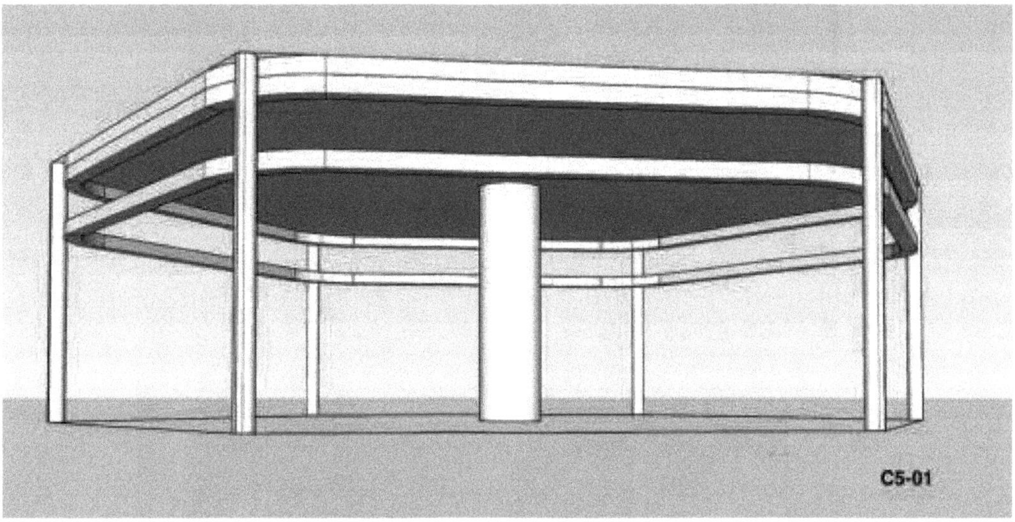


The 7230 Honeycomb module (C5-01) is based on a hexagonal prism with 7.23m (23ft-9in) sides and 4.0m (13'-1 ½") height. In addition to the central support, there is a round column at each corner. I used this module, which I created at the early stage of my work on elevated urbanism, for the following proposal.

To illustrate the potential for development over an environmentally sensitive area, I have selected Ayr, a small town of ten thousand people south of Kitchener, Ontario, which has magnificent natural areas threatened by urban sprawl (C5-02, C5-03 & C5-03a).

The raised street creates an open triangular space above the natural landscape and connects the train station with the business and/or residential towers. Thanks to its unique flexibility, the triangular grid offers countless building configuration opportunities and possible expansion.

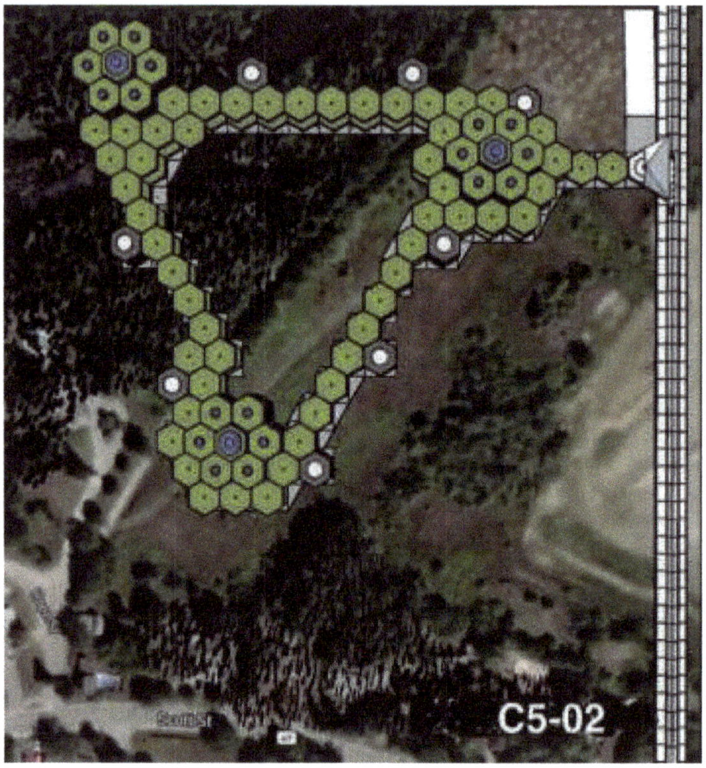


The layout can be determined to minimize the impact on the original landscape, primarily to avoid cutting trees.

See video: https://vimeo.com/673899571

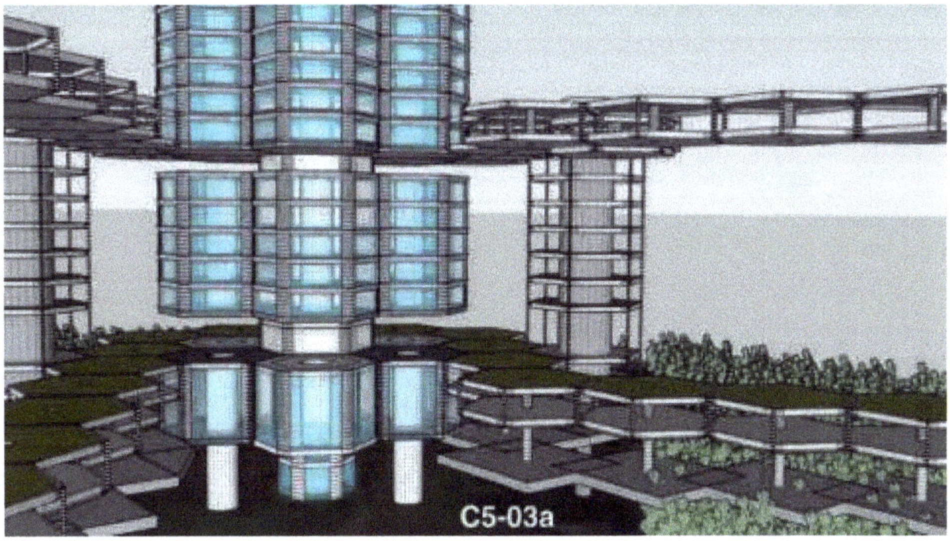


Above is a bird's view of the interior garden (C5-03a).

Next, I used hexagonal modules 2900, described in Chapter 4, for a concept of Honeycomb City raised above a natural area of woods and lakes north of Burlington, Ontario, which has several unique and innovative features (C5-04, C5-05, C5-05a, C5-06).

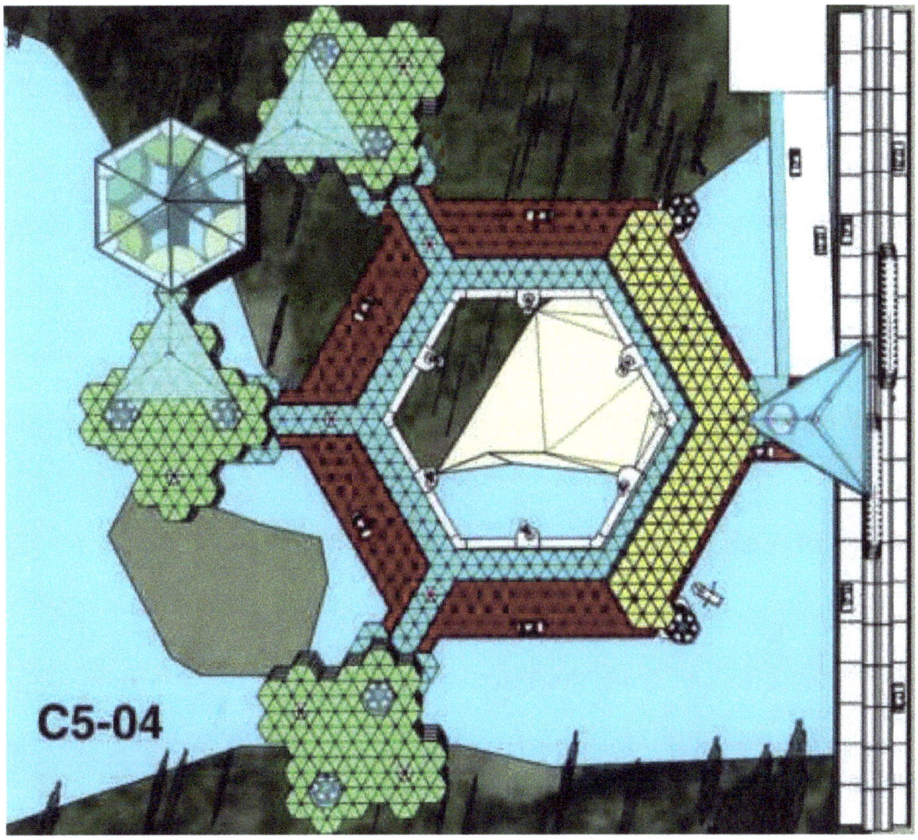

## Safe and convenient transportation

A network of raised fast trains and highways shall connect elevated urban communities. The walking distance to and from the train station shall not exceed 500m. The raised low-speed streets shall be used only by residents and emergency vehicles and are ideal for driverless electric cars, which would provide transportation to and from the train station.

As shown in image C5-04, the one-way street going around the recreational area provides access to each residential and/or business building and the one-storey commercial space located in front of the train station. The pedestrian areas are depicted in cyan colour, and the vehicular areas with maroon (C5-09).

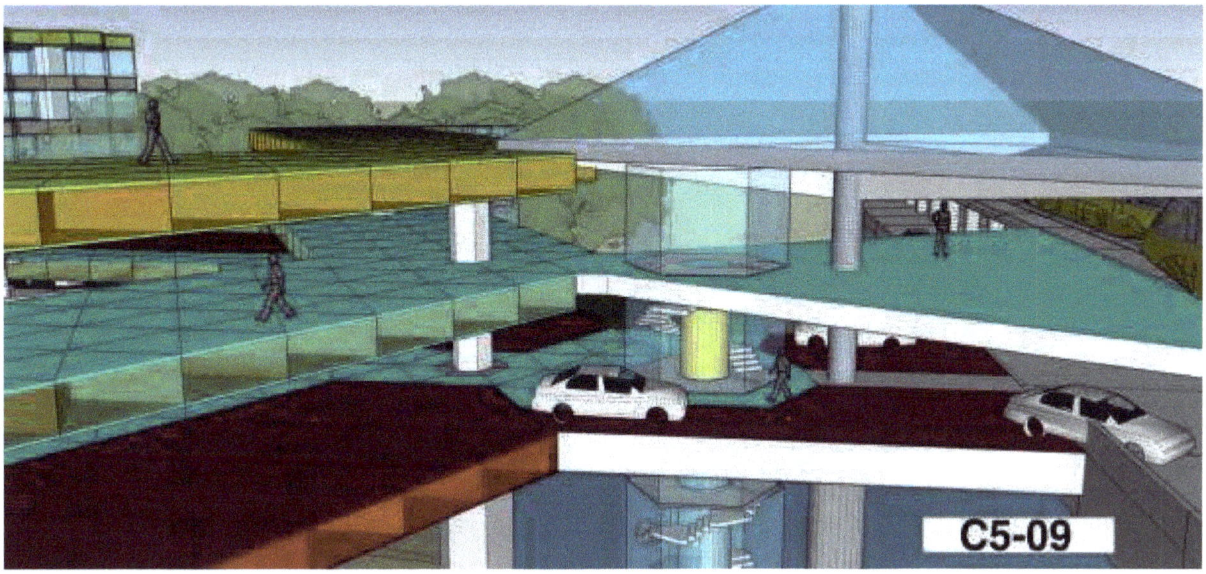


## Wheelchair Access Greenhouses

As a municipal building code official, I have, for long, struggled with the lack of provision in the building code for an emergency exit for wheelchairs in high-rise buildings.

Although exit stairs are required, people physically challenged, such as those confined to a wheelchair, the visually impaired, and the elderly, need to be evacuated by firemen. The justification had been that barrier-free ramps would occupy large areas to access all floors.

For a floor-to-floor height of 4.0m, six ramps are required to follow the perimeter of a hexagon with 14.50m sides, with start and finish landings corresponding to each floor. I established the dimensions of the hexagonal base of the greenhouse to fit into the module 2900 grid, thus creating flexibility in locating the hexagonal prism and connecting it with other buildings of the development as needed.

Each landing is expanded inside the hexagonal prism to form planters, depicted in the following image with various colours, and the entire structure is enveloped with glass to create a multilevel greenhouse (C5-07, C5-07a).

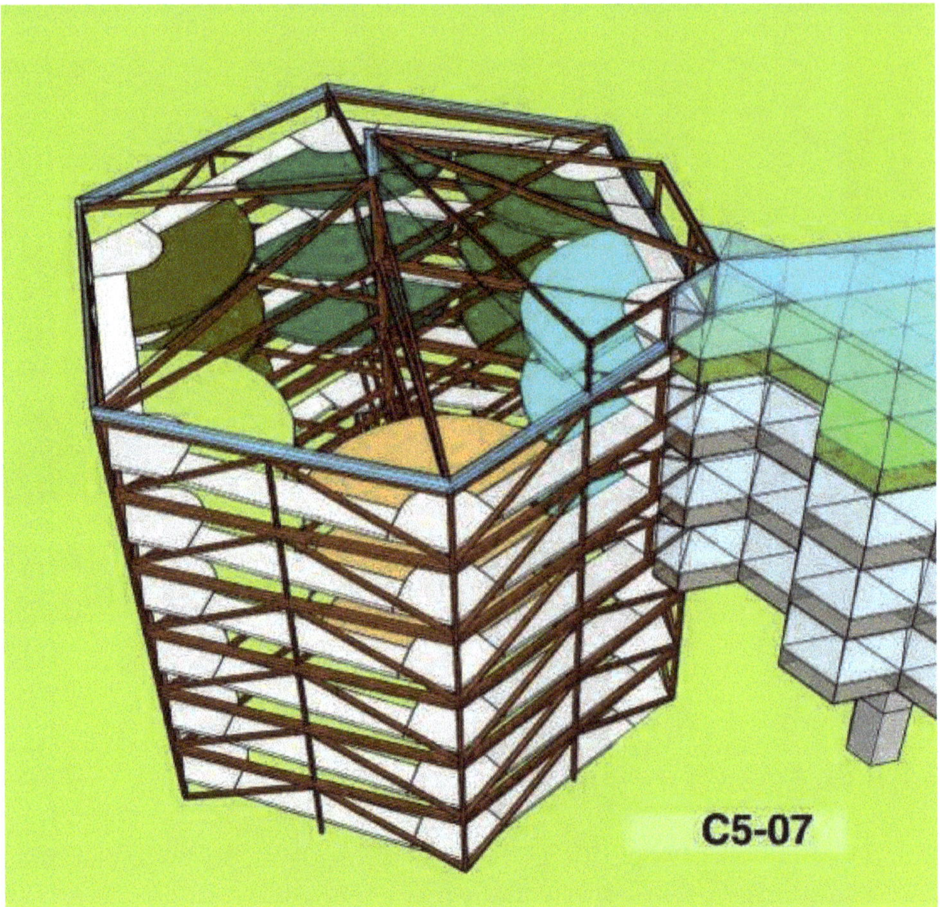

C5-07

C5-07a

## Access to waterways

One of the great advantages of an elevated urban community over a ground-based is that it can also be developed over lakes or rivers. The support piers, which enclose staircases, elevators, and vertical services, can be constructed in the water with minimal environmental disruption. This advantage brings several attractive benefits, including the use of boats (C5-08).

C5-08

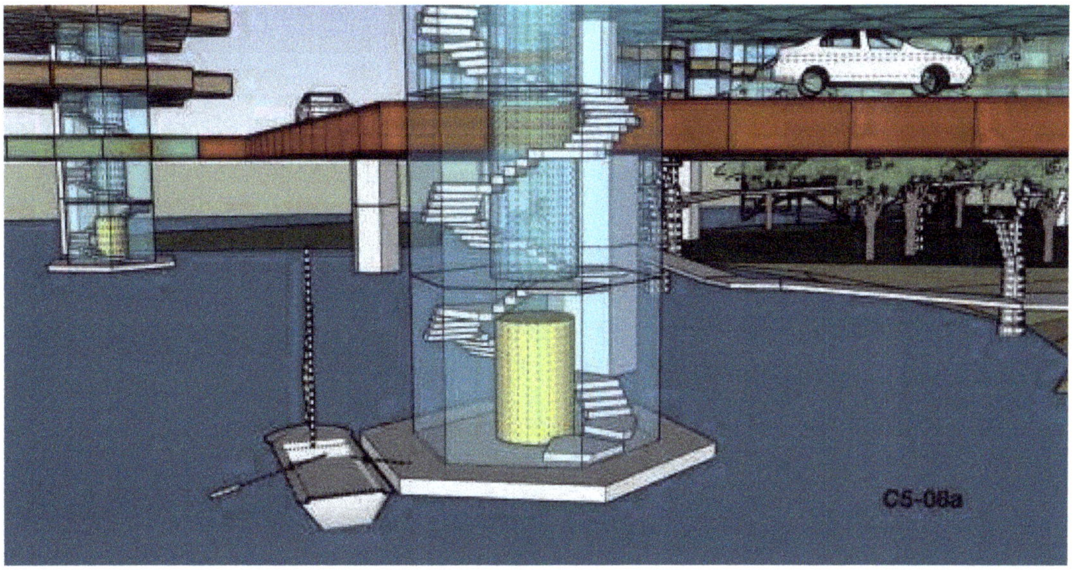

C5-08a

## Recreation

An artificial beach could be created, accessible to people with a physical challenge through ramps following the perimeter of the hexagonal recreational space. Sports facilities and playgrounds can be provided as well (C5-10, C5-11).

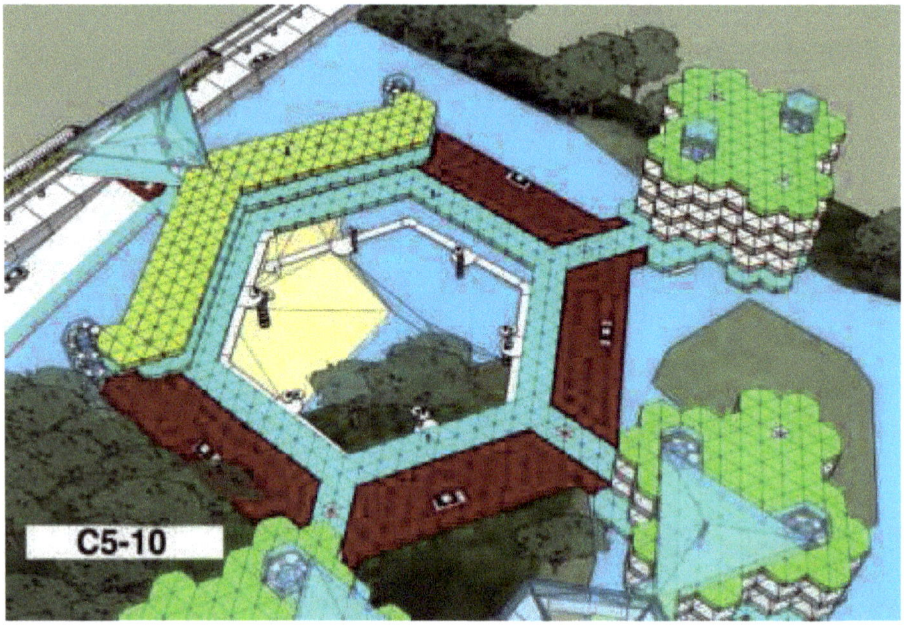

See video: https://vimeo.com/673327796

## The return of wildlife

Humans have a moral obligation to put an end to the destruction of wildlife habitats. One of the main contributors to this destruction is the fragmentation of natural areas throughout the world due to urbanization, forestry, farming, roads, railways, and power grids. Wild habitats have been reduced to shrinking islands, causing the continuous population to decrease and the extinction of many species.

Elevated urbanism could stop further fragmentation. It could even help wildlife to recover over time. One could only imagine what the positive results would be for both humans and animals. The French incorporate frog tunnels under their highways, while in Canada, warning signs of dear crossing are put up!

If allowed to safely move under the highways (C5-12, C5-14), many endangered species will recover. Bears, and other large predators, would prefer to search for food in the forest rather than adventure into our streets. An elevated urban community can be much easier protected and made safe for residents.

I am not advocating excluding people from the natural ground. It would be impossible to achieve anyway. However, the natural landscape under elevated urban communities can be made accessible in the same way as natural wildlife reserves are. We must accept individual responsibility for one's own safety and respect for the natural habitat.

# Chapter 6: Honeycomb in Space

Space exploration will lead to space migration. The lifespan of the planet is limited, and it depends on the life of the Sun. There have been approximate calculations made of how long the Sun would last before it burns all its hydrogen. Yet long before this happens, the Earth will probably be destroyed by some other natural cataclysm. As a most positive effect, the freedom of expanding human society into space will allow the earth to recover its natural glory.

Present activities that pose a threat to people and nature, such as hazardous industries, could be moved to earth-orbiting stations and to the moon. In 1923, the Romanian Hermann Oberth was the first to use the term "space station" for his wheel-like facility that would help launch astronauts to the moon and Mars. After World War II, German rocket scientist Wernher von Braun published his vision of a spinning wheel-shaped station, which was used in the 1968 film "2001: A Space Odyssey."

More recently, in 2018, Tom Marotta and Al Globus published the High Frontier: An Easier Way, which was inspired by Gerard O'Neill's 1976 book The High Frontier – Human Colonies in Space. They share good scientific discoveries of this century, including lighter radiation shielding required in Equatorial Low Earth Orbit (ELEO) and more flexible rotation health restrictions. Most importantly, they provide a theoretical path to cultivating a successful human community in ELEO.

Future discoveries in this century will open the universe to further human exploration, and the production of limitless fusion energy will make extraterrestrial travel convenient and affordable. Conquering gravity will probably be the greatest human achievement of all time and could free humanity from its prehistoric ties, just like cutting the umbilical cord allows a baby to develop as an independent individual.

Gravity is one of the four basic forces in Physics (Gravity, Electricity & Magnetism, and Nuclear Forces) and the most difficult to comprehend. The only clear thing is that it is associated with mass. There are no known present ways of replacing mass for the creation of gravity. However, there are ways to "fake" it.

In his theory of relativity, Einstein discovered through the equation E = mc 2 (energy = mass times the speed of light, squared) that mass is simply another form of energy.

So at least theoretically, it should be possible to create a large mass from a lot of energy, and this could open the door to creating artificial gravity. Applying this idea to a space station, each floor would incorporate electromagnetic devices distributed as required.

These devices would generate an electromagnetic field that replaces the Earth's gravitational pull. The net effect would be a level of artificial gravity that could be adjusted with the intensity of the electric current. This would require a very large quantity of energy in electric form, which may be obtained from the sun with photovoltaic cells.

However, the effect of various levels of electromagnetic fields on the human body has not been established yet.

There are studies that identify electromagnetism as "possibly cancer generating," but more research needs to be done. In the absence of a conclusive answer, adaptation to weightlessness and microgravity may be a safer, less risky alternative. In this respect, the experience acquired by the astronauts who lived in the International Space Station for extended periods indicates that the effects of microgravity on human health are manageable. Nevertheless, even low levels of gravity would make life easier, especially for extended periods of time.

Therefore, the floors could be treated to act as magnets. A promising avenue for achieving this objective is to add magnetic compounds into the composition of flooring materials, furniture, and most objects used in everyday life, even shoes and clothing. A good example is Carbon nanofoam, discovered in 1997 by Andrei Rode at the Australian National University. This allotrope of Carbon has a low-density cluster of atoms. It is a poor electrical conductor and contains numerous unpaired electrons, which gives it an unusual but very useful property: it is attracted to magnets.

A more realistic artificial gravity could be created by centripetal forces generated by spinning a space station on a large radius circle.

I have been interested in extra-terrestrial expansion since 2010, and I conceived a concept for a space habitat called 3STAR (C6-01), which I included in my book "Brainstorming for a Beautiful Planet" (ISBN 9781484851333).

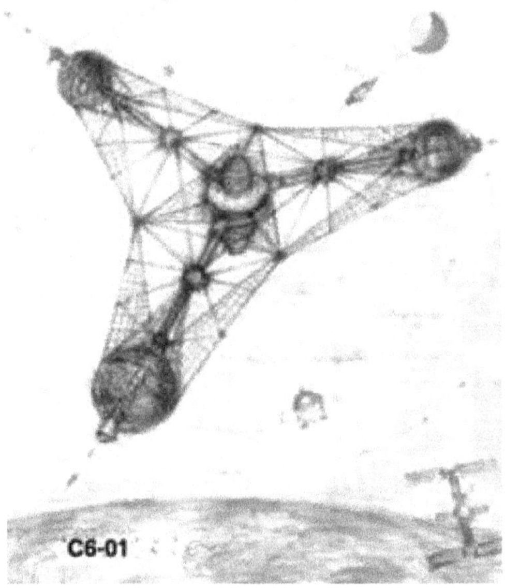


Several designs for space habitats have been proposed in the last century, such as the Stanford torus (1-mile diameter donut-shaped), the Bernal sphere (16 Km diameter), the O'Neill cylinder (3 Km radius), and combinations of them. In essence, they all create artificial gravity with centripetal force produced by spinning the entire space station. Unfortunately, they have the disadvantage of being very large and inflexible in the sense that they could not serve their purpose before being fully completed. This poses tremendous economic and technological difficulties.

3STAR is a relatively small-scale space prototype that could be technically and economically feasible.

Any earth-orbiting space habitat could be developed in stages, using the International Space Station as a platform. This will make good use of the great achievements of ISS and would ensure continuity in space exploration.

Having the flexibility to conduct multi-disciplinary research is crucial for any such prototype. 3STAR would provide independent modules for facilitating parallel experiments in various scientific fields, permaculture, energy self-sufficiency, simulated gravity, space architecture, low and near-zero gravity manufacturing, etc.

3STAR provides this opportunity thanks to its configuration that not only allows development in stages but also invites expansion.

3STAR could also provide a platform for travel to Mars and other planets and for establishing a lunar base, avoiding Earth's gravity. Its telescope could observe the universe better than anything that could ever be built on Earth.

The Space Honeycomb continues the 3STAR's concept of creating artificial gravity. Compared to other proposals for spaceships that provide artificial gravity, the Honeycomb Space Traveler shall be more feasible and cost-effective. Most of its components are hexagonal modules, which have superior structural performance and can be manufactured with 3D printing.

The modules could be sent to LEO with rockets and assembled there to form a fully functional extraterrestrial habitat that could also travel to other planets. The flexibility of the honeycomb modular system allows many possible configurations which can be expanded. The three peripheral habitat blocks shown in this model, which revolve around a 500M trajectory, are hollow to create interior gardens for a livable artificial environment.

The modules are configured based on the following design criteria needed for space habitat and travel:

- the structure and envelope to be prefabricated and assembled into one building block, which could be sent in rockets into ELEO (Equatorial Low Earth Orbit) for assembly.
- to facilitate stacking in and unstacking from a rocket for transportation to ELEO.
- to allow easy transportation from the factory to the launching site by truck, train, or ship.
- to permit easy assembly in ELEO, with drones and robots.
- to be installed only with bolted connections, without welding.
- to provide flexibility in space habitat configuration.
- to be demountable and replaceable with new modules.
- to be long-lasting.
- to be reusable.

The module presented here (C6-2) is a hexagonal prism with 7.2m (23.62') sides and 3.0m (9.84') height. This module is assembled from six triangular prisms with 7.2m (23.62') equal sides and a radius of 8.32m (27.3'). The author assumed that this triangular prism could be stacked in the cargo space of a rocket with an exterior diameter of 9.0m (29.53'), such as the SpaceX Cargo. At least 10 modules could be stacked in the cargo space.

The Honeycomb Space Traveler illustrated in this chapter (C6-04) has 3973 hexagonal modules, each one using six triangular modules, which means that 23,838 triangular modules are required to construct the entire spaceship, as shown. This translates into 2384 rocket launches. It seems a lot. However, assembling a smaller functional Honeycomb Space Traveler in the first phase would require a fraction of this number. Modules can be added or relocated as needed. This is the magic of the honeycomb!

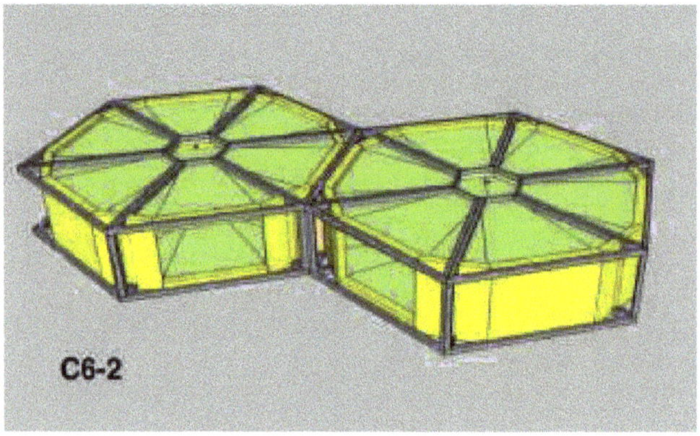


The envelope must be designed to resist the harsh space temperatures, which are as cold as minus 250 degrees Fahrenheit to as hot as 250 degrees in the sunlight.

To contain the interior air pressure, the interior corners of the hexagon must be rounded, and the space between the interior and exterior skins requires special insulation. The external hexagonal prism has open edges with exposed structural framing, which allows access for bolting.

I propose two different concepts for the Honeycomb Space Station. HSS-A (C6-04) is a follow-up to 3STAR, which is described in the 1st edition of Brainstorming for a Beautiful Planet. I called it 3STAR because it has three geodesic spheres for habitation, which rotate around a central hub and are connected to it by three spokes. HSS-A replaces the geodesic spheres with habitation blocks constructed with honeycomb modules.

Following is a view of the Honeycomb Space Station (HSS-A6), with six habitation blocks, six converging spokes, and six peripheral circulation links (C6-03). The hexagonal grid permits expansion in a multitude of directions. The central hub has two main parts, a static hub and a dynamic hub, which revolves around the static part. The converging spokes are connected to the dynamic hub. Near the intersection with the peripheral links, each spoke has a sleeve of additional modules for the rotation propulsion system.

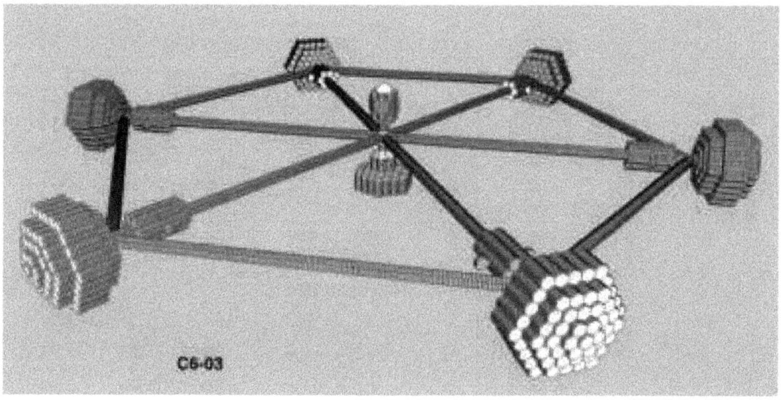


See video: https://vimeo.com/697081752

An expanded static hub could accommodate two assemblies of habitation blocks, spokes, and peripheral links, with the two assemblies would rotate in opposite directions for stability in orbit (C6-03a).

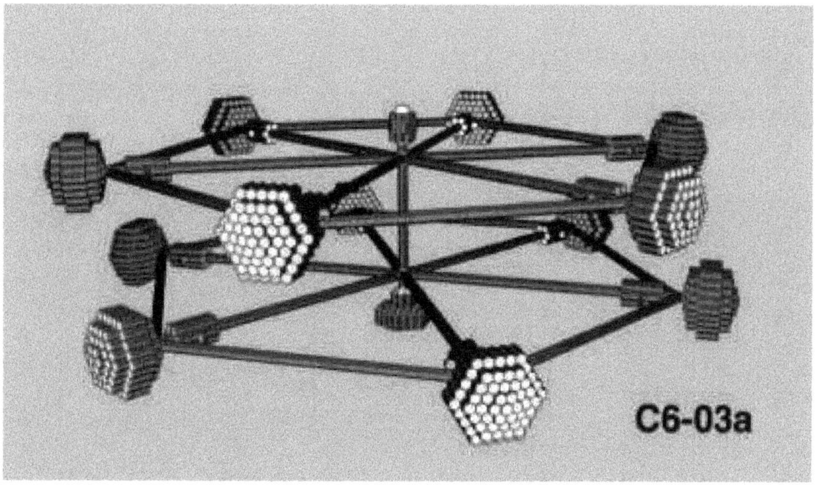

## The Honeycomb Space Traveler

The Honeycomb Space Traveler (HST), or HSS-A3 (C6-04), is a variation of Honeycomb Space Station HSS-A, adapted for long-term space travel. The three habitation blocks with artificial gravity provide living quarters for pilots and their families. They are connected to the hub with three circulation spokes, which also incorporate storage and lab rooms. The extended static hub is provided with docking modules and storage. It also has the command pilot's cabin at one end and the propulsion engines at the other.

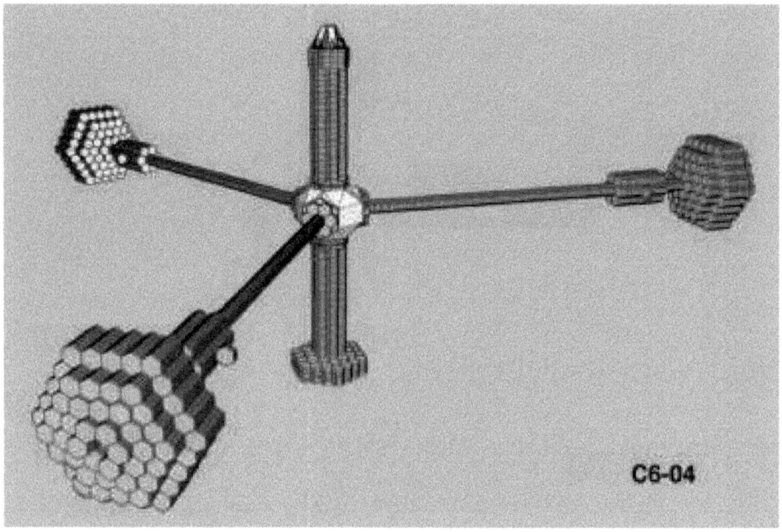


See video: https://vimeo.com/697083242

Following is a transparent exterior view of an HSS-A habitation block (C6-05) connected to the intersection of a spoke with two peripheral links.

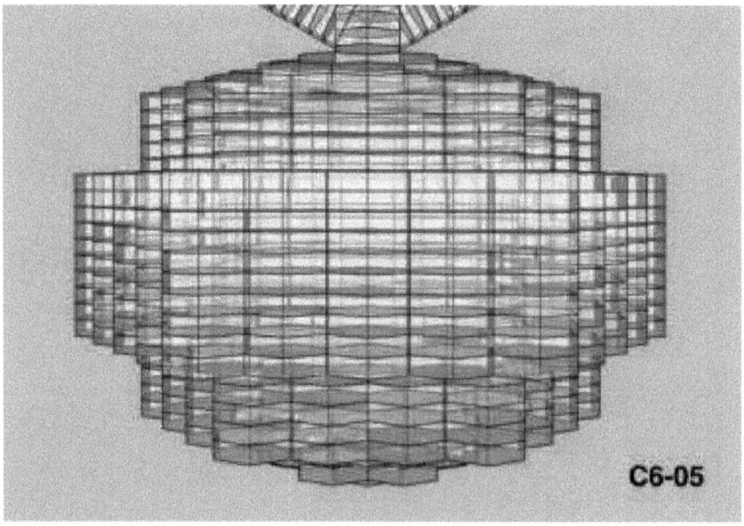

## Living in Space

The sections through a habitation block (C6-06 & C6-07) show the interior garden, which receives light from the top module connected to the spoke and peripheral links. The habitation modules open onto the garden, which has its soil deposited at the highest gravity level; therefore, it will attract all dust and moisture from the air. Over time, the amount of topsoil will grow thanks to natural cycles.

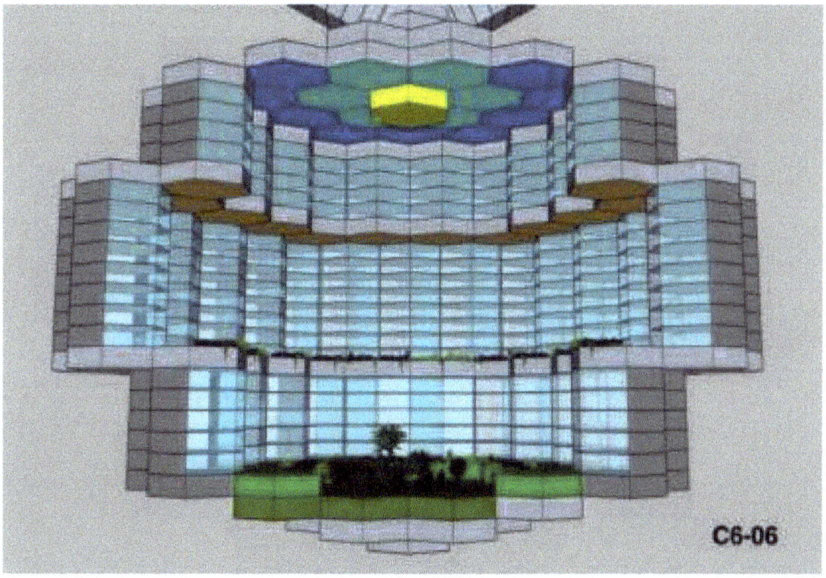


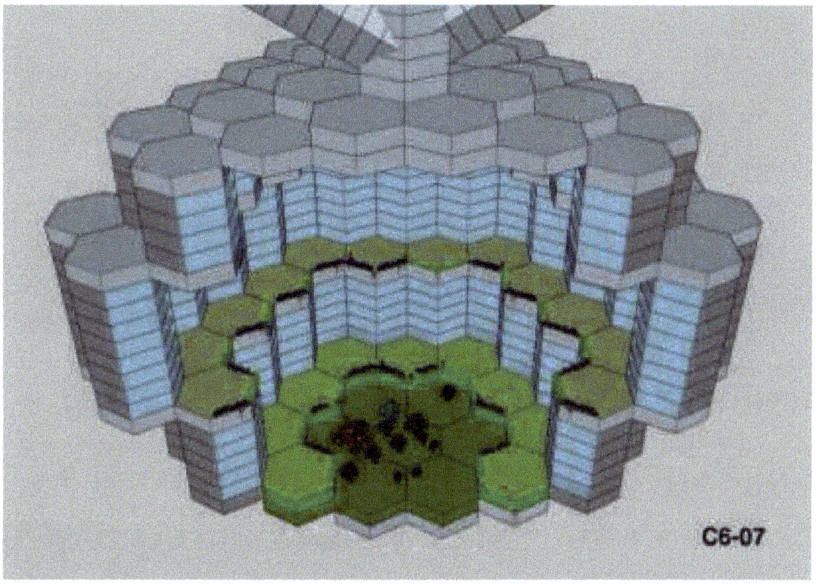


See video: https://vimeo.com/697084332

## The Rotation Hub

The rotation hub is designed to permit circulation between the dynamic and static assemblies. This is achieved through concentric rings, with the external rings revolving slowly around the internal rings of the static assembly. The narrow gap between the static and dynamic rings is filled with a lubricating material and covered with an adjustable threshold.

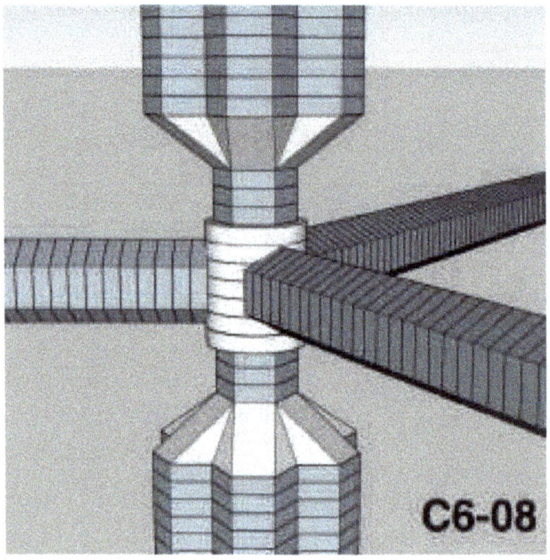


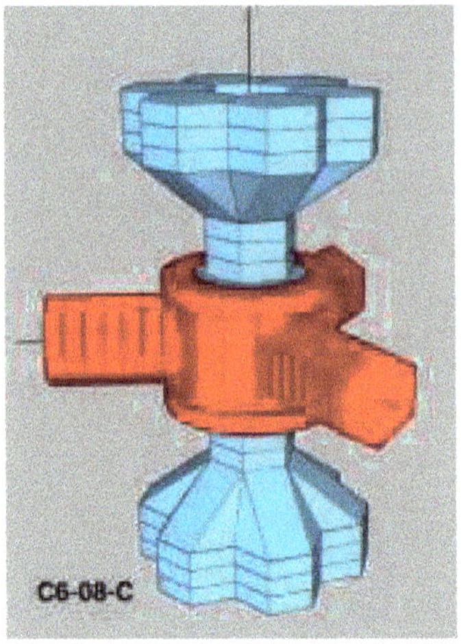


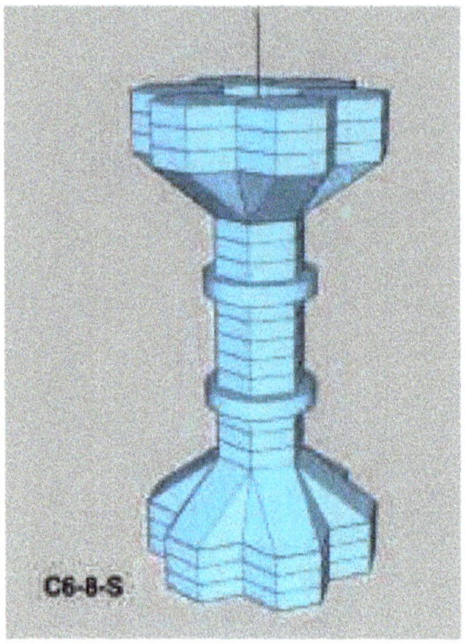


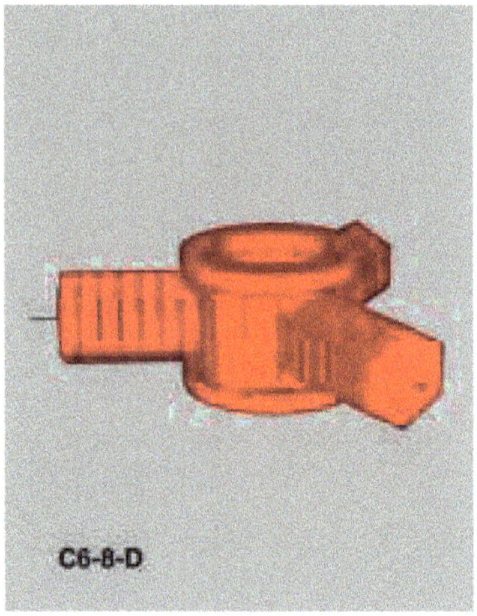


## The Honeycomb Space Station B

Inspired by the ideas of Gerrard O'Neill, I designed the Honeycomb Space Station B (HSS-B) concept. The first alternative is Honeycomb Space Station X (HSS-X) (C6-09), which includes two cylinders of 500m radius, revolving in opposite directions for stability.

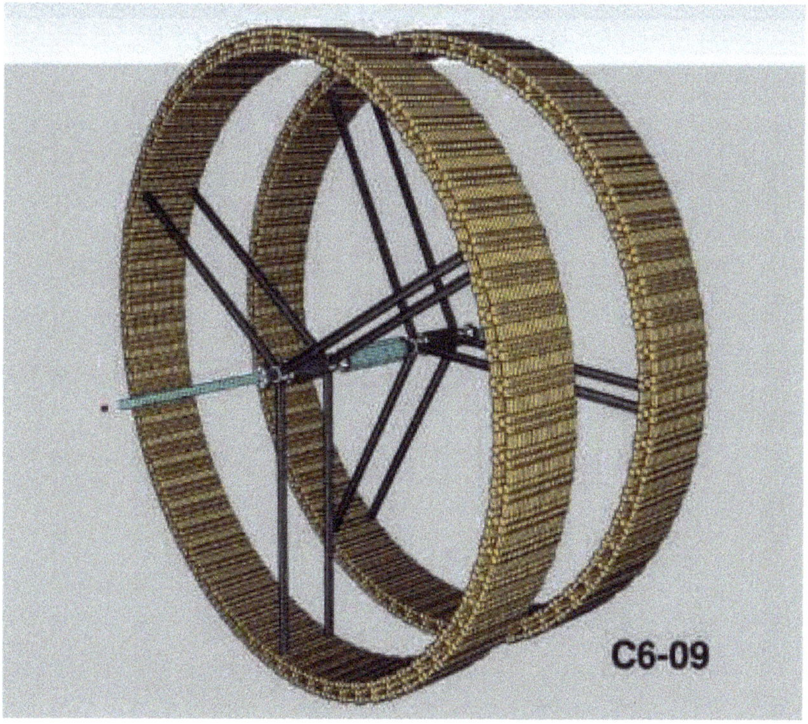


See video: https://vimeo.com/712570034

These cylinders are assembled from two different modules: a hexagonal prism with 7.2m (22.62') sides and 6.0m (19.69') height (C6-2) and a trapezoidal prism with 6.0m (19.69') height and two different bases. The following image illustrates three hexagonal modules (C6-10).

The smaller trapezoid prism has 6.0m (19.69') height, 17.84m (58.53') & 18.46m (60.56') bases. The larger trapezoid prism has 6.0m (19.69') height, 19.73m (64.73') & 20.35m (66.77') bases (C6-11).

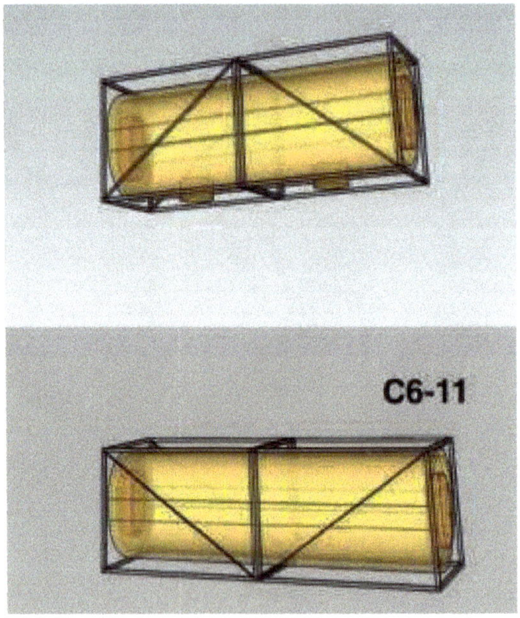


The trapezoid prism modules are distributed on two concentric orbits. The smaller trapezoid prism modules are placed on the smaller orbit with a radius of 495.37m (1625.23'). The larger trapezoid prism modules are placed on the larger orbit with a radius of 516.96m (1696.06'). Each trapezoid prism module is placed between two hexagonal prism modules, which form part of two groups of five hexagonal prism modules. The rotation circle is divided into 72 sectors with an angle of 5°. This translates into an inclination angle of 2.5° between a trapezoid prism module and the adjacent hexagonal prism modules.

Each module has an external structural framing, which envelopes an insulated cylinder. The framing is provided with bolt connectors which permits the modules to be assembled in space by robots. They also facilitate expansion. The space between the framing and cylinder accommodates cables and pipes, which can be inspected and maintained from the outside.

All modules are made with two separate components, the structural framing on the outside and the insulated cylindrical capsule on the inside. The two components shall be pre-assembled prior to launching to space, including all service pipes and cables.

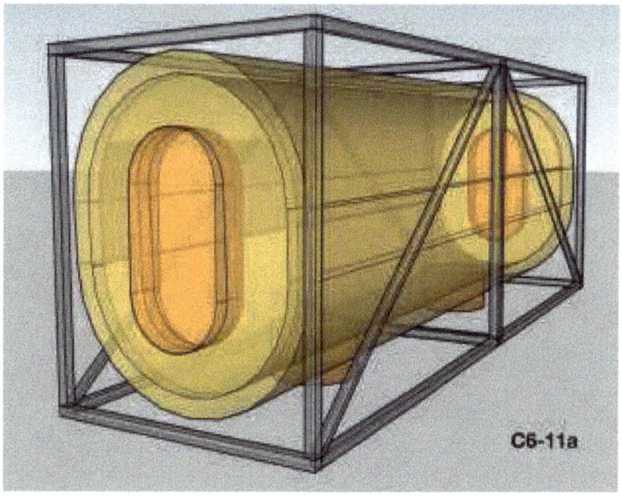


The structural framing is pre-assembled from elements manufactured from high-strength, lightweight metal alloys or carbon fiber. As shown in the illustrations, most of the structural framing is open and is accessible to bolting by robotic drones, which shall perform all the installation of the modules in ELEO.

The insulated capsule shall be prefabricated, possibly by 3D printing. The exterior skin shall be impact and radiation resistant. The penetrations required for service pipes and cables shall be configured to exclude any leakage of the interior conditioned air, which will attempt to escape to the space vacuum. The technology is already in place, as demonstrated by the ISS.

The doors connecting the modules are part of an insulated cylindrical enclosure that penetrates through the skins of the capsule and are incorporated at the time of manufacturing. This communication assembly is provided with various types of airlocks as required by location. Such airlocks have already been tested on the ISS.

The honeycomb structure provides superior flexibility in three dimensions for placing the modules and communicating between them. This flexibility also results in the easier and faster installation of the modules.

Each cylindrical capsule will be designed to accommodate a specific activity and will be provided with doors and windows, depending on its location. The space created by six hexagonal and two trapezoid modules receives sunlight and can be used for farming, recreation, and other outdoor activities.

The following illustration shows a 24m courtyard formed by four layers of modules (C6-12).

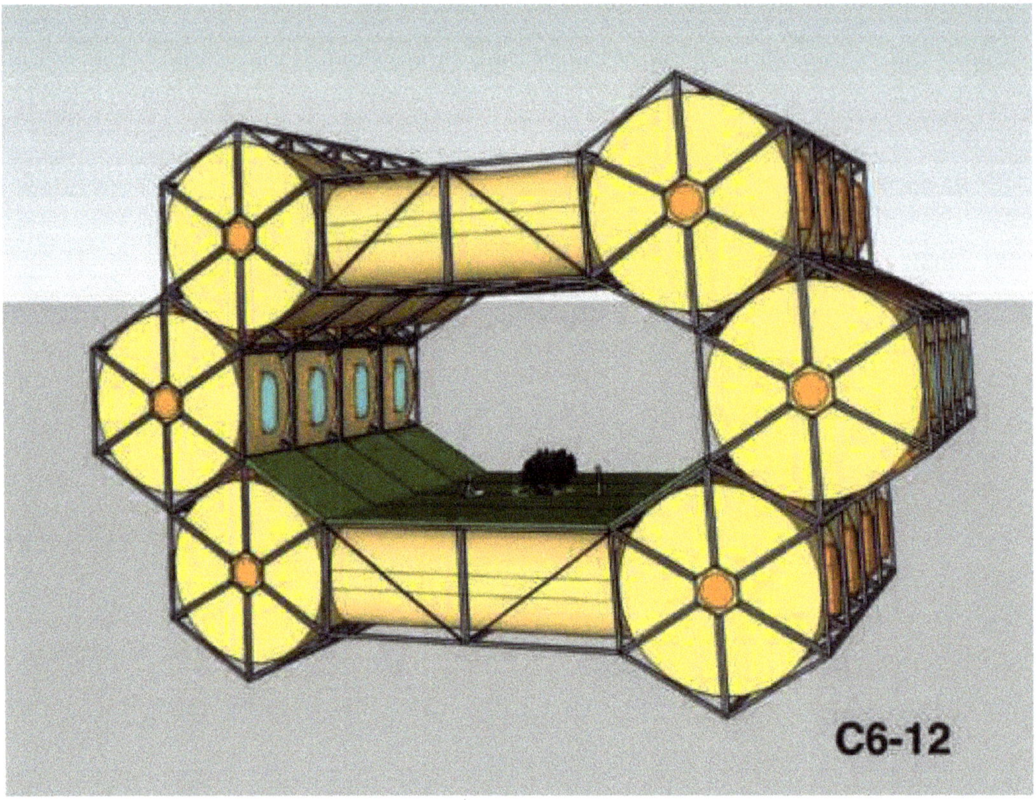

See video: https://vimeo.com/712961167

After all, modules are installed, the walls, floor, and ceiling of the courtyard are enclosed and sealed with an airtight membrane supported by the module framing. This will permit the creation of a microclimate with artificial air produced by equipment installed in the trapezoid modules. These modules will also serve to generate energy, water, and all other materials necessary for the functioning of the space station.

Each cylinder for the Honeycomb Space Station X illustrated earlier (C6-09) has twenty module layers of 6m each, to a total of 120m (393.70'), but it can be reduced or enlarged. It has 14,588 habitation modules. Each such module contains a cylindrical space with an exterior volume of 626.14m3 and an interior volume of 467.51m3, the difference being taken by the shells and insulation. The interior circular floor area is 97.4m2 (1048.44s.f). Assuming an occupancy of at least two people per module, this space station could accommodate approximately 30,000 people.

**A proposed strategy of development and construction by space agencies in cooperation with private space companies.**

The first step is design development, which includes multidisciplinary research, architectural & interior design, and modeling, multidisciplinary engineering (structural, mechanical, astronautical, energetical, etc.). Concomitant with design development, testing of all components shall be done in laboratories, simulated space environments, and near ISS.

The second step, concomitant with design development and testing, is the development of the vehicle for delivery to ELEO and the robotic equipment for installing the modules. The vehicle of delivery could be one of the current reusable rockets, which would need to be adapted to accommodate the HSM modules.

The third step is the construction of a small mock-up assembly in ELEO to test the installation of the modules in space and their performance over several years with astronauts on board.

The fourth step is the construction of the HSS-X, which would be done in the following stages:

1. Two-layer wheel with an outer radius of approx. 520m. One layer has 360 hexagonal modules and 144 trapezoidal modules. Each of the three spokes has 471 hexagonal modules. The hub has a static block in the centre of a dynamic block connected to the spokes.

2. Activate rotation engines located in the outer hexagonal modules in tandem, and test clockwise rotation. Test station maneuvering engines are located at both ends of the static hub.

3. Enclosing courtyards and testing counter-clock rotation.

4. Adding wheels (including spokes), extending hub, enclosing courtyards, and testing clockwise rotation.

5. Repeat items 1, 2, 3, & 4 on the opposite side, but with opposite rotation.

6. Enclose the four end-wheels with space structures supporting glass panels to create two cylinders, as shown in the next image, C6-13.

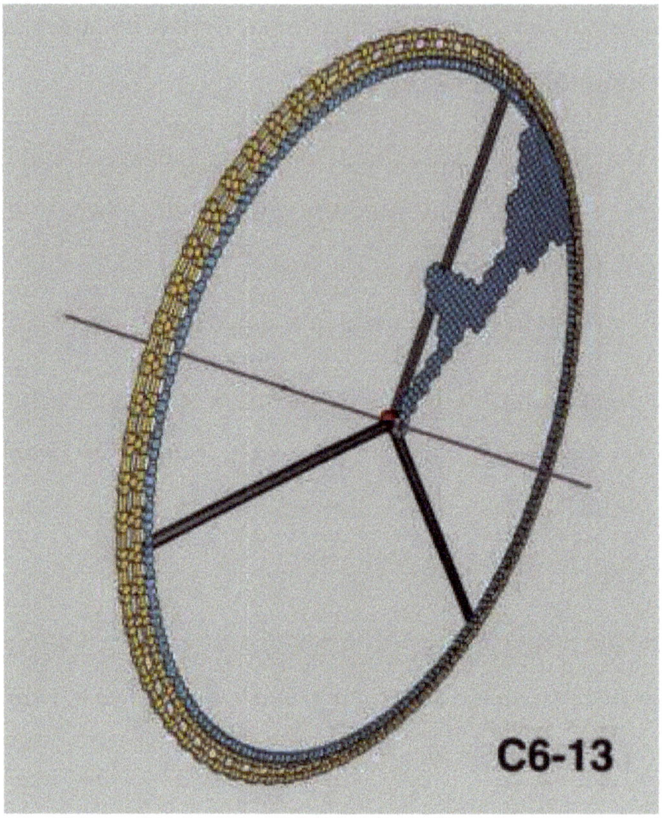


The space station could be expanded by increasing the length of the cylinders. Furthermore, the end bases of the cylinders could also be enclosed with a space structure supporting glass panels, which would create a large space with an artificial atmosphere. The scaling up of the habitat would make life more pleasant for the residents. Image C6-13 illustrates the installation of the glass panels (blue) in progress.

## Connection Details

The three-dimensional hexagonal structure of each module is constructed with hollow sections. These incorporate rods that can slide into the hollow sections of adjacent modules (C6-14).

## Viability of the Honeycomb in Space

Due to their unique configuration, the honeycomb space modules are most appropriate for creating commercially viable, flexible, and expandable space habitats with artificial gravity.

Following are the steps required for developing an extraterrestrial power station and hotel.

### Phase 1

- Costing
- Multidisciplinary design and engineering of the modules.
- Design of AI systems, space installation robots, and space manufacturing equipment.
- Testing of components in the laboratory.
- Testing of modules in the desert and arctic environments.
- Promotion and Marketing.
- Financing

### Phase 2

- Manufacturing the starter station (ST) with 8 modules, including a PV system and EmDrive electric propulsion, for testing in space.
- Manufacturing of installation robots.
- Fabricating the equipment for space manufacturing.
- Launching into LEO.
- Assembling the 8 starter modules.
- Testing of all systems, EmDrive electric propulsion, maneuvering, orientation, connections, PV system, etc.
- Receiving a shipment of regolith from the Moon or Asteroids.
- Testing of equipment for space manufacturing.

- Testing of PV energy transmission to the receptor on Earth.
- Multidisciplinary design and engineering adjustments as required.
- Preparing the starter station (ST) for space tourism.

In their book "The High Frontier: An Easier Way," Tom Marotta and Al Globus stress the importance of introducing a commercial factor in outer space developments due to their high cost. Governments alone, through taxpayers' contributions, cannot be expected to cover the entire investment required, which amounts to many billions of dollars. Therefore, procuring revenues from space projects is essential. The main sources of these revenues are extra-terrestrial tourism, energy generation, and mining.

The image below (C6-15) illustrates phase 2, the starter station (ST) for a space hotel in ELEO.

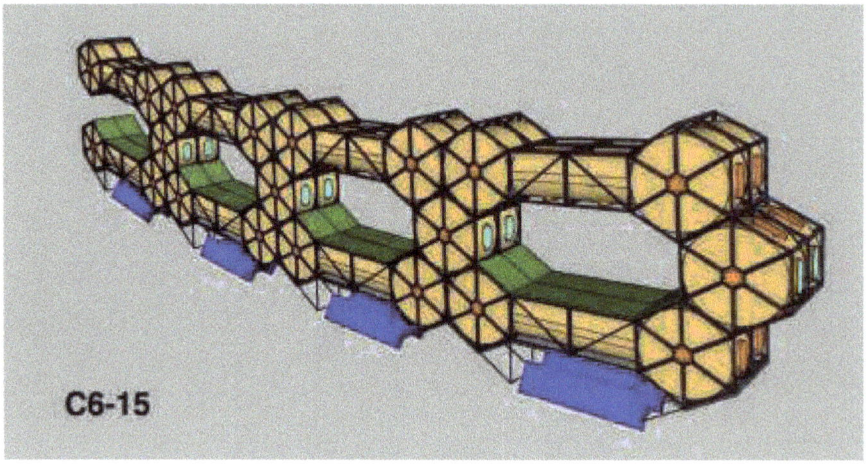


See video: https://vimeo.com/724653459

## Phase 3

- Manufacturing of some module components in and near the starter station (ST).
- Manufacturing of some module components on Earth, as required.
- Expanding the starter station (ST) along axis X to create a six-layer station of 48 modules.
- Expanding the starter station (ST) around the centre to create a 500m radius wheel of 864 modules.

- Installation of the 3 spokes with 471 modules per spoke.
- Installation of the Static and Dynamic Hubs at the centre.

Following is the schematic for a larger space station HSS-2X (C6-16), which implements some of the elements of the cylinders designed by Gerard K. O'Neill.

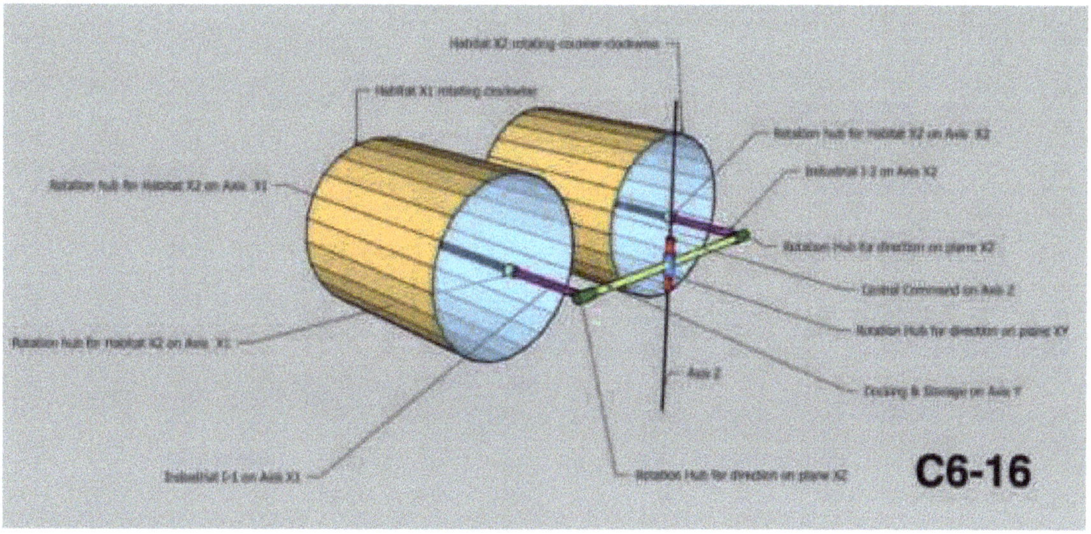


See video: https://vimeo.com/719739222

## Comparison between Honeycomb Space Stations-A and B.

- In model A, each habitation block is self-sufficient, with the interior garden producing food for its residents. On the other hand, in model B, a large "green cylinder" can be created to feed the entire space station.
- Although both models A and B have emergency evacuation shuttles for returning to Earth or to a Moon base in case of a space catastrophe occurring far in the solar system, at least one habitation block of model A could be designed to bring all people home. This consideration could be the determining factor in the decision to travel and live in space.
- Model A requires less investment for creating a starter station with recreation and open space to make a long-term stay more pleasant. It may be the pioneer station needed to build the larger model B, which could fulfil Gerard O'Neill's dream for large space communities with a life comparable to Earth's.

# Epilogue

Are you wondering about the miracle happening in the image below? It is the triumph of Nature.

She has been telling us for thousands of years: “Build the way I do, so you may survive.” But we didn’t listen and now are facing dire consequences.

The choices we have are clear: Reduce our footprint on Earth and expand human society to outer space. If our planet is destroyed, all people will perish, regardless of their wealth. Our instinct of survival should tell us that we must make sacrifices today to ensure we will have a tomorrow.

My ideas and concepts presented in this book are neither the ultimate solution nor exclusive of other ideas and concepts. Some of them have already been proven by experience or common sense. Others require further research, multidisciplinary design, and testing. So, there is still a lot of work to do before they can be implemented, and this work needs many brains, eyes, and hands. I hope this book is helping you to consider participating in any way you can.

# List of Illustrations



**Foreword**

**Chapter 3: Flexsola City**



**Chapter 4: The Honeycomb**

**Chapter 5: Honeycomb City**

**Chapter 5: Honeycomb in Space**



**Epilogue**

www.ingramcontent.com/pod-product-compliance
Lightning Source LLC
LaVergne TN
LVHW070855210926
848482LV00016B/941
* 9 7 8 1 9 1 6 5 4 0 5 0 7 *